PRODUCTION SCHEDULING

Edited by **Rodrigo da Rosa Righi**

Production Scheduling
Edited by Rodrigo da Rosa Righi

Published by InTech
Janeza Trdine 9, 51000 Rijeka, Croatia

Publishing Process Manager Martina Blecic

Technical Editor Teodora Smiljanic

Cover Designer InTech Design Team

Additional hard copies can be obtained from orders@intechopen.com

Production Scheduling, Edited by Rodrigo da Rosa Righi
p. cm.
ISBN 978-953-307-935-6

For EU product safety concerns:
IN TECH d.o.o., Prolaz Marije Krucifikse Kozulić 3, 51000 Rijeka, Croatia,
info@intechopen.com or visit our website at intechopen.com.

Meet the editor

Dr. Rodrigo da Rosa Righi is a professor and a researcher at the University of Vale do Rio dos Sinos (Unisinos) in the South of Brazil. He is also a member of the Applied Computing Graduate Program at the same university, advising graduate and undergraduate students. He received his BS degree in Computer Science from the Federal University of Santa Maria (UFSM), Brazil, in 2003. Following that, Dr Righi obtained his MS and PhD degrees in Computer Science from the Federal University of Rio Grande do Sul (UFRGS), Brazil, in 2005 and 2009, respectively. During his doctorate studies, Dr Righi performed a "sandwich" stage at Tecnische Universitaet Berlin, Germany. Currently, his research interests include performance analysis, production planning and scheduling, process migration, load balancing and resource provisioning on Cluster, Grid and Cloud environments. Moreover, he is an expert on embedded computing systems which employ Linux or WinCE operating systems. He is also a member of the IEEE Computer Society and ACM.

Contents

Preface

Generally speaking, scheduling is the procedure of mapping a set of tasks or jobs (studied objects) to a set of target resources efficiently. It is an important tool in manufacturing and engineering, where it can have a major impact on the productivity of a process. More specifically, as a part of a larger planning and scheduling process, production scheduling is essential for the proper functioning of a manufacturing enterprise. Its primary goal is to maximize throughput and optimize resource utilization, while at the same time minimizing operating costs. Moreover, production scheduling is a challenging field of research in manufacturing engineering, with a high degree of exposure to the evolution of both global industries and information technology.

In manufacturing, the purpose of production scheduling is to minimize production time and costs, by telling a production facility when to make something, with which staff, and using which equipment. Furthermore, advances in computer science have resulted in a wide selection of software capable of undertaking constraint-based optimization, modeling, what-if scenarios and experiment design. Current challenges in production scheduling deal with the impact of globalization, the usage of cutting-edge technologies for improved decision-making and novel modeling perspectives, including cognitive and social aspects of production scheduling.

Basically, the benefits of production scheduling include: process change-over reduction; inventory reduction, leveling; reduced scheduling effort; increased production efficiency; labor load leveling; accurate delivery date quotes; real time information and so on. Considering the benefits mentioned above, the current book aims to gather information on the latest advances concerning the development of the state-of-the-art techniques for production scheduling.

I am extremely proud to present to you ten book chapters divided into five sections. Section 1 shows some rescheduling strategies, policies, and methods for production scheduling. Section 2 presents two chapters about flow shop scheduling. Section 3 describes heuristic and metaheuristic methods for solving the scheduling problem in an efficient manner. In addition, two test cases are presented in Section 4. The first uses simulation, while the second shows a real implementation of a production scheduling system. Section 5 presents some modeling strategies for building production scheduling systems.

We can observe that the topic of scheduling has a wide impact in several areas. Consequentially, the content of this book is not limited to engineering, but it also touches on other areas of knowledge, like biological, chemical and computational fields. This book will therefore be of interest to those working in the decision-making branches of production, in various operational research areas, as well as computational method design. People from diverse backgrounds, academia, industry and research, can take advantage of this volume. Finally, I would like to point out that I would much appreciate if you sent your comments about the book to me at rrrighi@unisinos.br.

Prof. Dr. Rodrigo da Rosa Righi
Professor and researcher in the Applied Computing Graduate Program (PIPCA)
Universidade do Vale do Rio dos Sinos (UNISINOS)
São Leopoldo - Brazil

Part 1

Rescheduling Strategies, Policies and Methods

1

Process Rescheduling in High Performance Computing Environments

Rodrigo da Rosa Righi and Lucas Graebin
Programa Interdisciplinar de Pós-Graduação em Computação Aplicada
Universidade do Vale do Rio dos Sinos
Brazil

1. Introduction

Scheduling is an important tool for manufacturing and engineering, where it can have a major impact on the productivity of a process (Min et al., 2009). In manufacturing, the purpose of production scheduling is to minimize the production time and costs, by informing a production facility when to make, with which staff, and on which equipment (Zhu et al., 2011). Nowadays, it is possible to observe the use of specific computational tools for production scheduling in which greatly outperform older manual scheduling methods. These tools implement mathematical programming methods that model the problem as an optimization issue where some objective, *e.g.* total duration, must be minimized (or yet maximized) (Yao & Zhu, 2010).

The concepts behind manufacturing and production scheduling can be employed in serie of contexts where optimization field takes place (Delias et al., 2011; Fan, 2011; Wang et al., 2011). In a common meaning, scheduling consists in formulating plans in which organize objects for operating efficiently in a specific context. Especially, this chapter discusses about a rescheduling method for high-performance environments (HPC) like Computational Grids, or just Grids (Yu & Buyya, 2005). The presented method deals with processes, in which represent an execution entity of the operating system. Here, we named scheduling as the first mapping of processes to resources in the Grid. Thus, the method is responsible for process rescheduling, so changing the first process-processors assignment in the distribute system by offering efficient scheduling plans along the application's lifetime. As an optimization problem , its main purpose focuses on minimizing the execution time of the application as a whole.

As grid computing emerged and got widely used, resources of multiple clusters became the dominant computing nodes of the grid (El Kabbany et al., 2011; Qin et al., 2010; Xhafa & Abraham, 2010). Applications containing routines for solving linear systems and fast Fourier transform (FFT) are typical examples of tightly-coupled parallel applications that may take profit from the power of cluster-of-clusters (Sanjay & Vadhiyar, 2009). Given that these clusters can be heterogeneous and the links among them are normally not fast, scheduling and load balancing are two key techniques that must be organized to reach high performance in this architecture. An alternative for offering this treatment focuses on scheduling using process migration. Therefore, we can reshape the process-resource matching at runtime in accordance with both the behavior of the processes and the resources (processors and network).

Generally, process migration is implemented within the application, resulting in a close coupling between the application and the algorithms' data structures. Such an implementation is not extensible, due to the specificity of the shared data structure. Even more, some initiatives use explicit calls in the application code (Bhandarkar et al., 2000) and obligate extra executions to get tuned scheduling data (Silva et al., 2005; Yang & Chou, 2009). A different migration approach happens at middleware level, where changes in the application code and previous knowledge about the system are usually not required. Considering this, we have developed a process rescheduling model called MigBSP (da Rosa Righi et al., 2010). It was designed to work with phases-based applications with BSP behavior (Bulk Synchronous Parallel) and acts over cluster-of-clusters architectures. MigBSP extensively uses heuristics to adapt the interval between migration calls, to analyze the behavior regularity of each process as well as to select the candidates for migration. Heuristics were employed since the problem of finding the optimum scheduling in heterogeneous system is in general NP-hard (Xhafa & Abraham, 2010).

Concerning the choosing of the processes, MigBSP creates a priority list based on the highest Potential of Migration (*PM*) of each process (da Rosa Righi et al., 2010). *PM* combines the migration costs with data from both computation and communication phases in order to create an unified scheduling metric. Using a hierarchy notion based on two levels (Goldchleger et al., 2004), each *PM* element concentrates a target process and a specific destination site. Figure 1 goes through the *PM* approach. The process denoted in the top of the mentioned list was always selected to be inspected for migration viability. Although we achieved good results when using this approach, we agree that an optimized one can deal with multiple processes when rescheduling verification takes place. A possibility could concern the selection of a percentage of processes based on the highest *PM*. Nevertheless, a question arises: How can one reach an optimized percentage value for dynamic applications and heterogeneous environments? A solution could involve the testing of several hand-tuned parameter instances and the comparison of the results. Certainly, this idea is time consuming and new applications and resources require a new series of tests.

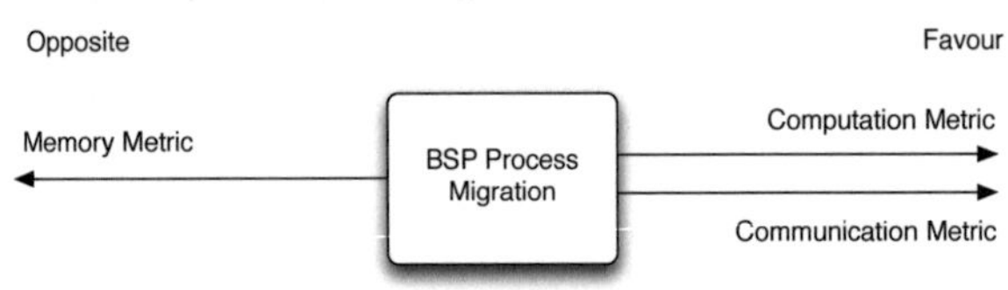

Fig. 1. Potential of Migration (*PM*) combines Computation, Communication and Memory data in order to offer an unified scheduling metric. The rationale of creating *PM* consists in evaluating migrations of processes to different sites, reducing the number of tests in the rescheduling moment.

After developing the first version of MigBSP, we have observed the promotion of intelligent scheduling systems which adjust their parameters on the fly and hide intrinsic complexity and optimization decisions from users (Ding et al., 2009; Nascimento et al., 2007; Sanjay & Vadhiyar, 2009). In this context, we developed a new heuristic named **AutoMig** that selects one or more candidates for migration automatically. We took advantage of both List Scheduling (Duselis et al., 2009) and Backtracking (Baritompa et al., 2007) concepts to evaluate the migration impact on each element of the *PM* list in an autonomous fashion. In addition, other AutoMig's strength comprises the needlessness to complete an additional MigBSP parameter for getting more than one migratable process on rescheduling activation. The scheduling evaluation uses a prediction function (*pf*) that considers the migration costs and works following the concept of a BSP superstep (Bonorden, 2007). The lowest forecast value indicates the most profitable plan for process rescheduling.

This book chapter aims to describe AutoMig in details. Particularly, we evaluated it by using a BSP application that computes image compression based on the Fractal method (Guo et al., 2009). Considering that the programmer does not need to change his/her application nor add a parameter on rescheduling model, the results with migration were satisfactory and totaled a mean gain of 7.9%. Furthermore, this classification is due to fact that AutoMig does not know any application and resource descriptions in advance. The results showed a serie of situations where AutoMig outperforms the heuristic that elects only one process. Next section shows MigBSP briefly and serves as the basis for understanding the proposed heuristic.

2. MigBSP: Rescheduling model

MigBSP answers the following issues regarding load balancing: (i) "When" to launch the migration; (ii) "Which" processes are candidates for migration; (iii) "Where" to put an elected process. In a previous paper we described the ideas to treat these questions in details (da Rosa Righi et al., 2010). The model requires both unicast and asynchronous communications among the processes. The target architecture is heterogeneous and composed by clusters, supercomputers and/or local networks. The heterogeneity issue considers the processors' clock speed (all processors have the same machine architecture), as well as network speed and level (Fast and Gigabit Ethernet and cluster-of-clusters environments, for instance). Such an architecture is assembled with abstractions of Sets (different sites) and Set Managers. As an example, a specific Set could be composed by the nodes from a cluster. Set Managers are responsible for scheduling, capturing data from a specific Set and exchanging it among other managers.

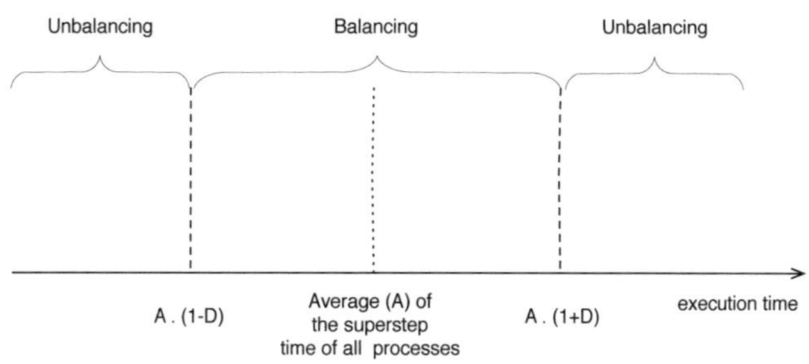

Fig. 2. Analysis of balancing and unbalancing situations which depend on the distance *D* from the average time *A*.

The decision for process remapping is taken at the end of a superstep. Aiming to generate the least intrusiveness in application as possible, we applied two adaptations that control the value of α ($\alpha \in \mathbb{N}^*$). α is updated at each rescheduling call and will indicate the interval for the next one. To store the variations on system state, a temporary variable called α' is used and updated at each superstep through the increment or decrement of one unit. The adaptations' objectives are: (i) to postpone the rescheduling call if the processes are balanced or to turn it more frequent, otherwise; (ii) to delay this call if a pattern without migrations on ω past calls is observed. A variable denoted D is used to indicate a percentage of how far the slowest and the fastest processes may be from the average to consider the processes balanced. In summary, the higher the value of α, the lower the model's impact on application runtime.

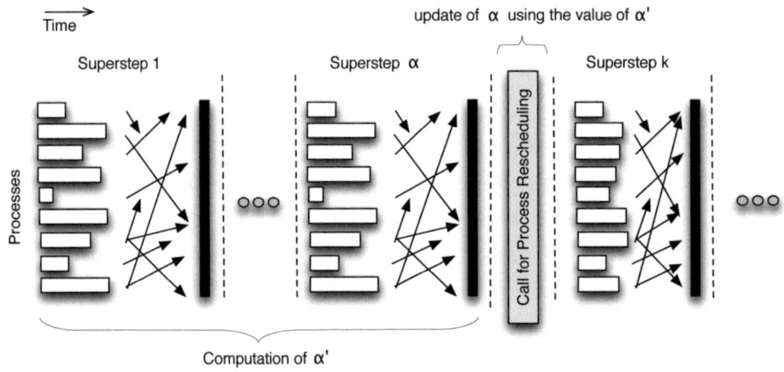

Fig. 3. Overview of an application execution with MigBSP: α parameter indicates the next interval for process rescheduling and depends on the value of α', which is updated at each BSP superstep.

The balanced state is based on the superstep time of each BSP process. Figure 2 depicts how both balanced and unbalanced situations can be reached. In implementation view, the processes save their superstep time in a vector and pass it to their Set Managers when rescheduling is activated. Following this, all Set Managers exchange their information. Set Managers have the times of each BSP process and compute the balancing situation. Therefore, each manager knows the α' variation locally. Figure 3 illustrates an example of the interaction between a BSP application and MigBSP. As we can observe, this figure presents the expected result when calling rescheduling actions since a reduction in time can be verified in the remaining BSP supersteps.

The answer for "Which" is solved through our decision function called Potential of Migration (*PM*). Each process i computes n functions $PM(i,j)$, where n is the number of Sets and j means a Set. The key rationale consists in performing only a subset of the processes-resources tests at the rescheduling moment. Considering that Grid scheduling is multi-objective in its general formulation (Xhafa & Abraham, 2010), $PM(i,j)$ is found using Computation, Communication and Memory metrics as we can see in Equation 1. The relation among them is based on the notion of force from Physics. Computation and Communication act in favor of migration, while Memory works in an opposite direction. The greater the value of $PM(i,j)$, the more prone the BSP processes will be to migrate.

$$PM(i,j) = Comp(i,j) + Comm(i,j) - Mem(i,j) \tag{1}$$

The Computation metric — $Comp(i,j)$ — uses a Computation Pattern $P_{\text{comp}}(i)$ that measures the stability of a process i regarding the number of instructions at each superstep (see Equation 2). This value is close to 1 if the process is regular and close to 0 otherwise. Other element in $Comp(i,j)$ is a computation time prediction $CTP_{k+\alpha-1}(i)$ of the process i at superstep $k+\alpha-1$ (last superstep executed before process rescheduling). Supposing that $CT_t(i)$ is the computation time of the process i during superstep t, then the prediction $CTP_{k+\alpha-1}(i)$ uses the Aging concept as follows(da Rosa Righi et al., 2010; Tanenbaum, 2003).

$$CTP_t(i) = \begin{cases} CT_t(i) & if\ t = k \\ \frac{1}{2}CTP_{t-1}(i) + \frac{1}{2}CT_t(i) & if k < t \leq k + \alpha - 1 \end{cases}$$

$Comp(i,j)$ also presents an index $ISet_{k+\alpha-1}$(j). This index informs the average capacity of Set j at the $k + \alpha - 1^{th}$ superstep. For each processor in a Set, its load is multiplied by its theoretical capacity. Concerning this, the Set Managers compute a performance average of their Sets and exchange this value. Each manager calculates $ISet(j)$ for each Set normalizing their performance average by its own average. In the sequence, all Set Managers pass $ISet(j)$ index to the BSP processes under their jurisdiction.

$$Comp(i,j) = P_{comp}(i) \, . \, CTP_{k+\alpha-1}(i) \, . \, ISet_{k+\alpha-1}(j) \tag{2}$$

In the same way, the Communication metric — $Comm(i,j)$ — computes the Communication Pattern $P_{\text{comm}}(i,j)$ between processes and Sets (see Equation 3). Furthermore, this metric uses a communication time prediction $BTP_{k+\alpha-1}(i,j)$ involving the process i and Set j between two rebalancing activations. This last parameter employs the same idea used to compute $CTP\_t(i)$, where the prediction value is more strongly influenced by recent supersteps. The result of Equation 3 increases if the process i has a regularity considering the received bytes from processes of Set j and performs slower communication actions to this Set.

$$Comm(i,j) = P_{comm}(i,j) \, . \, BTP_{k+\alpha-1} \tag{3}$$

The Memory metric — $Mem(i,j)$ — composition can be seen in Equation 4. Firstly, the memory space in bytes of considered process is captured through $M(i)$. After that, the transfer time of 1 byte to the destination Set is calculated by $T(i,j)$ function. The communication involving process i is established with the Set Manager of each considered Set. Finally, the time spent on migration operations of process i to Set j is calculated through $Mig(i,j)$ function. These operations are dependent of operating system, as well as the tool used for providing process migration.

$$Mem(i,j) = M(i) \, . \, T(i,j) \; + \; Mig(i,j) \tag{4}$$

BSP processes calculate $PM(i,j)$ locally. At each rescheduling call, each process passes its highest $PM(i,j)$ to its Set Manager. This last entity exchanges the PM of its processes with other managers. As mentioned earlier, there is a heuristic to choose the candidate for migration which is based on a decreasing-sorted list composed by the highest PM value of each process. This heuristic chooses the head of the list. $PM(i,j)$ of a candidate process i is associated to a Set j intrinsically. The manager of this Set will select the most suitable processor to receive the process i.

Before any migration, its viability is verified considering the following data: (i) the external load on source and destination processors; (ii) the BSP processes that both processors are executing; (iii) the simulation of considered process running on destination processor; (iv) the time of communication actions considering local and destination processors; (v) migration costs. Then, we computed two times: t_1 and t_2. t_1 means the local execution of process i, while t_2 encompasses its execution on the other processor and includes the costs. For each candidate, a new resource is chosen if $t_1 > t_2$.

3. AutoMig: A novel heuristic to select the suitable processes for migration

AutoMig's self-organizes the migratable processes without programmer intervention. It can elect not only one but a collection of processes at the rescheduling moment. Especially, AutoMig's proposal solves the problem described below.

- **Problem Statement** - Given n BSP processes and a list of the highest *PM* (Potential of Migration) of each one at the rescheduling moment, the challenge consists in creating and evaluating at maximum n new scheduling plans and to choose the most profitable one among those that outperform the current processes-resources mapping.

AutoMig solves this question by using the concepts from List Scheduling and Backtracking. Firstly, we sort the *PM* list in a decreasing-ordered manner. Thus, the tests begin by the process on the head since its rescheduling represents better chances of migration gains. Secondly, AutoMig proposes n scheduling attempts (where n is the number of processes) by incrementing the movement of only one process at each new plan. This idea is based on the Backtracking functioning, where each partial candidate is the parent of candidates that differ from it by a single extension step. Figure 4 depicts an example of this approach, where a single migration on level l causes an impact on $l+1$. For instance, the performance forecast for process "A" in the third *PM* considers its own migration and the fact that "E" and "B" were migrated previously. Algorithm 1 presents AutoMig's approach in details.

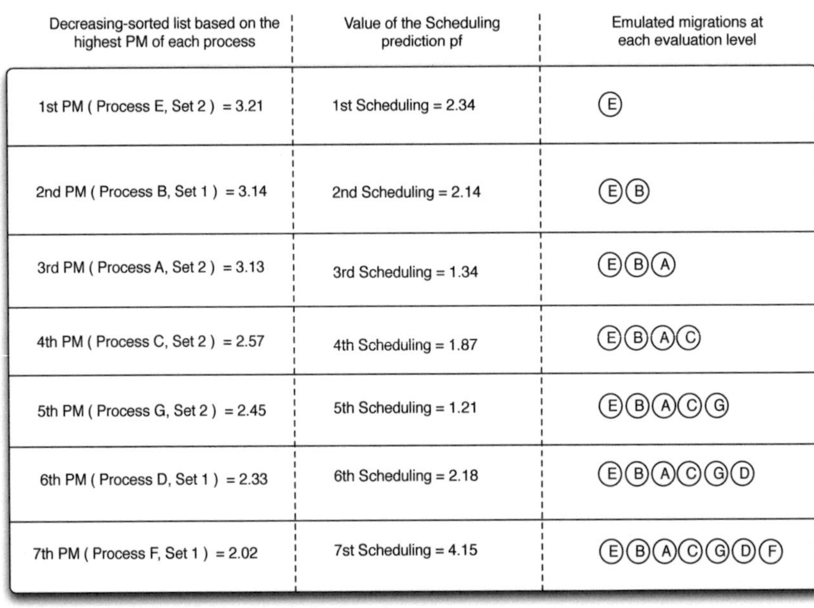

Fig. 4. Example of the AutoMig's approach. Only one process is migrated at each level of the *PM* list. A migration of a process on level l presents an impact in $l+1$ and so on.

The main part of AutoMig concerns its prediction function *pf*. *pf* emulates the time of a superstep by analyzing the computation and communication parts of the processes. Both parts are computed through Equations 5 and 6, respectively. They work with data collected at the superstep before calling the rescheduling facility. In addition, *pf* considers information about the migration costs of the processes to the Sets. The final selection of migratable processes is obtained through verifying the lowest *pf*. The processes in the level belonging to this prediction are elected for migration if their rescheduling outperforms the *pf* for the current mapping.

At the rescheduling call, each process passes the following data to its manager: (*i*) its highest *PM*; (*ii*) a vector with its migration costs (*Mem* metric) for each Set; (*iii*) the number of instructions; (*iv*) a vector which contains the number of bytes involved on communication actions to each Set. Each manager exchanges *PM* values and uses them to create a decreasing-sorted list. Task 5 of Algorithm 1 is responsible for getting data to evaluate the scheduling of the current mapping.

At each level of the *PM* list, the data of the target process is transferred to the destination Set. For instance, data from process 'E' is transferred to Set 2 according to the example illustrated in Figure 4. Thus, the manager on the destination Set will choose a suitable processor for the process and will calculate Equations 5 and 6 for it. Aiming to minimize multicast communication among the managers at each *pf* computation, each Set Manager computes $Time_p$ and $Comm_p$ for the processes under its jurisdiction and save the results together with the specific level of the list. After performing the tasks for each element on *PM* list, the managers exchange their vectors and compute *pf* for each level of the list as well as for the present scheduling (task 12 in Algorithm 1).

Equation 5 computes $Time_p(i)$, where i means a specific process. $Time_p(i)$ uses data related to the computing power and the load of the processor in which process i executes currently or is being tested for rescheduling. $cpu\_load(i)$ represents the CPU load average on the last 15 minutes. This time interval was adopted based on work of Vozmediano and Conde (Moreno-Vozmediano & Alonso-Conde, 2005). Equation 6 presents how we get the maximum communication time when considering process i and Set j. In this context, Set j may be the current Set of process i or a Set in which this process is being evaluated for migration. $T(k,j)$ refers to the transferring rate of 1 byte from the Set Manager of Set j to other Set Manager. $Bytes(i,k)$ works with the number of bytes transferred through the network among process i and all process belonging to Set k. Lastly, $Mig\_Costs(i,j)$ denotes the migration costs related to the sending of process i to Set j. It receives the value of the Mem metric, which also considers a process i and a Set j.

$$Time_p(i) = \frac{Instruction(i)}{(1 - cpu\_load(i)).cpu(i)} \tag{5}$$

$$\begin{aligned} Comm_p(i,j) = {} & Max_k \; (\; \forall \, k \in Sets \\ & (Bytes(i,k) \; . \; T(k,j)) \;) \end{aligned} \tag{6}$$

Algorithm 1 AutoMig's approach for selecting the processes

1: Each process computes *PM* locally (see Equation 1).
2: Each process passes its highest *PM*, together with the number of instructions and a vector that describes its communication actions, to the Set Manager.
3: Set Managers exchange *PM* data of their processes.
4: Set Managers create a sorted list based on the *PM* values with n elements (n is the number of processes).
5: Set Managers compute Equations 5 and 6 for their processes. The results will be used later for measuring the performance of the current mapping. Migrations costs are not considered.
6: **for** each element from 0 up to $n-1$ in the *PM* list **do**
7:     Considered element is analyzed. Set Manager of process i sends data about it to the Set Manager of Set j. The algorithm proceeds its calculus by considering that process i is passed to Set j.
8:     The manager on the destination Set chooses a suitable processor to receive the candidate process i.
9:     Set Managers compute Equations 5 and 6 for their processes.
10:     Set Managers save the results in a vector with the specific level of the *PM* list.
11: **end for**
12: Set Managers exchange data and compute *pf* for the current scheduling as well as for each level on *PM* list.
13: **if** $Min(pf)$ in the *PM* list < current *pf* **then**
14:     Considering the *PM* list, the processes in the level where *pf* was reached are selected for migration.
15:     Managers notify their elected processes to migrate.
16: **else**
17:     Migrations do not take place.
18: **end if**

$$pf = Max_i\,(Time_p(i)) + Max_{i,j}\,(Comm_p(i,j)) + Max_{i,j}\,(Mig\_Costs(i,j)) \tag{7}$$

Considering Equation 7, it is important to emphasize that each part may consider a different process i and Set j. For instance, a specific process may obtain the largest computation time, while other one expends more time in communication actions. Finally, AutoMig's selection approach uses a global strategy, where data from all processes are considered in the calculus. Normally, this strategy provides better results but requires synchronization points for capturing data. However, we take profit from the barriers of the BSP model for exchanging scheduling information, not paying an additional cost for that.

Kowk and Cheung (Kwok & Cheung, 2004) arranged the load balancing topic in four classes: (i) location policy; (ii) information policy; (iii) transfer policy and; (iv) selection policy. AutoMig answers the last issue by using a global strategy (Zaki et al., 1997). In this type of scheme, the decisions are made using a global knowledge, *i.e.*, data from all processes take part in the synchronization operation for processes replacement. The list of the highest *PM* of all BSP processes is known by Set Managers when attempting for migration. Therefore, the main advantage of global schemes comprises the better quality of load balancing decisions since the entire studied objects are considered. On the other hand, the synchronization is the most expensive part of this approach (El Kabbany et al., 2011; Zaki et al., 1997). However, we take profit from BSP model organization, which already imposes a barrier synchronization among the processes. Therefore, we do not need to pay an addition cost to use the global idea.

4. Evaluation methodology

We are simulating the functioning of a BSP-based Fractal Image Compression (FIC) application. FIC has generated much interest in the image compression community as competitor with well established techniques like JPEG and Wavelets (Guo et al., 2009). One of the main drawbacks of conventional FIC is the high encoding complexity whereas decoding time is much lower (Xing, 2008). Nevertheless, fractal coding offers promising performance in terms of image quality and compression ratios. Basically, FIC exploits similarities within images. These similarities are described by a contractive transformation of the image whose fixed point is close to the image itself. The image transformation consists of block transformation which approximate smaller parts of the image by larger ones. The smaller parts are called ranges and the larger ones domains. All ranges together form the image. The domains can be selected freely within the image.

For each range an appropriate domain must be found. A root mean-square-error (rms) distance is calculated in order to judge the quality of a single map. The encoding time depends on the number of domains whose each range must be compared to. A complete domain-poll of an image of size $t \times t$ with square domains of size $d \times d$ consists of $(t-d+1)^2$ domains. Furthermore, each domain has 8 isometries. So each range must be compared with $8(t-d+1)^2$ domains. The greater the number of domains, the better will be the compression quality. In addition, the application time increases as the number of domains increases as well.

Our BSP modeling considers the variation of both the range and domain sizes as well as the number of processes. Algorithm 2 presents the organization of a single superstep. Firstly, we are computing $\frac{t}{r}$ supersteps, where $t \times t$ is the image size and r is the size of square ranges. The goal is to compute a set of ranges at each superstep. For that, each superstep works over $\frac{t}{r}$ ranges since the image comprises a square. At each superstep, a range is computed against $8((\frac{t}{d})^2.\frac{1}{n})$ domains, where d represents the size of a domain and n the number of processes. Moreover, each process sends $\frac{t}{r}$ ranges before calling the barrier, which must be multiplied by 8 to find the number of bytes (we considered a range with 8 bytes in memory).

The main aim of the experimental evaluation is to observe the performance of MigBSP when working with AutoMig heuristic. Considering this, we applied simulation in three scenarios: (i) Application execution simply; (ii) Application execution with MigBSP scheduler without applying migrations; (iii) Application execution with MigBSP scheduler allowing migrations. Scenario ii consists in performing all scheduling calculus and decisions, but it does not comprise any migrations actually. Scenario iii adds the migrations costs on those processes that migrate from one processor to another. Both the BSP application and the model were developed using the SimGrid Simulator (MSG Module) (Casanova et al., 2008). It makes possible application modeling and processes migration. In addition, it is deterministic, where a specific input always results in the same output.

Algorithm 2 Modeling of a single superstep for the Fractal Image Compression problem

Considering a range-pool rp of the image ($0 \le rp \le \frac{t}{r} - 1$), where t and r mean the sides of the $t \times t$ image and $r \times r$ range, respectively
for each range in rp **do**
  for each domain belonging to specific process **do**
    for each isometry of a domain **do**
      calculate-rms(range, domain)
    end for
  end for
end for
Each process i ($0 \le i \le n - 1$) sends data to its right-neighbor $i + 1$. Process $n - 1$ sends data to process 0 (where n is the total number of processes)
Call for synchronization barrier

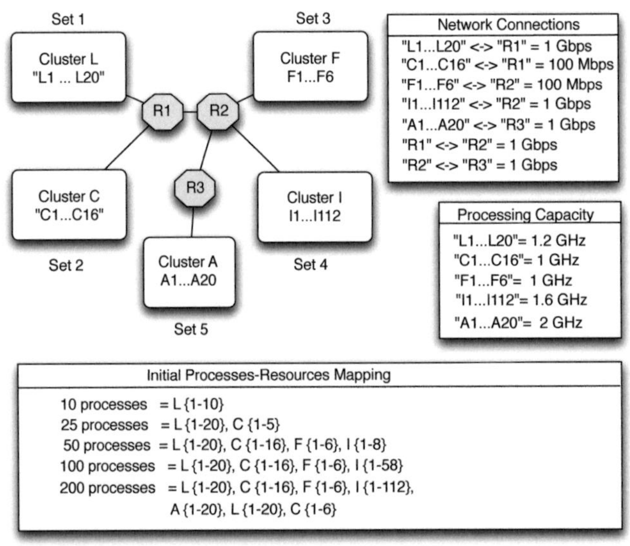

Fig. 5. Multiple Clusters-based topology, processing and network resources description and the initial processes-resources scheduling.

Aiming to test the scenarios, we assembled an infrastructure with five Sets which is depicted in Figure 5. A Set represents a cluster where each node has a single processor. The infrastructure permits us to analyze the impact of the heterogeneity issue on AutoMig's algorithms. Moreover, it represents the current resources at Unisinos University, Brazil. Initial tests were executed using α equal to 4 and D equal to 0.5. We observed the behavior of 10, 25, 50, 100 and 200 BSP processes. Their initial mapping to the resources may be viewed in Figure 5. Since the application proceeds in communications from process i to $i + 1$, we opted by using the contiguous approach in which a cluster is filled before passing to another one (Pascual et al., 2009). Furthermore, the values of 40, 20 and 10 were used for the side (d) of a square domain. The range value (r) is obtained by $\frac{d}{2}$. The considered figure is a square with side 1000. The lower the d variable, the greater the number of domains to be tested by each process. These parameters turn possible the verification of the AutoMig's overhead and situations where process rescheduling is applicable. Finally, the migration costs are based on previous executions with AMPI (Huang et al., 2006) on our clusters.

5. Analyzing AutoMig's overhead and decisions

Table 1 presents the initial tests when dealing with 40 and 20 for both domain and range sizes, respectively. This configuration enables a short number of domains to be computed by each process. Thus, the processes have a small computation grain and their migrations are not viable. *PM* values in all situations are negative, owing to the lower weight of the computation and communication actions if compared to the migration costs. Therefore, AutoMig figures out the lowest *pf* in which in reached through the current scheduling. Consequently, both times for scenario ii and iii are higher than the time spent in scenario i. In this context, a large overhead is imposed by MigBSP since the normal application execution is close to 1 second in average.

Processes	Scenario i	Scenario ii (MigBSP and AutoMig without Migrations)	Scenario iii (MigBSP and AutoMig enabling Migrations)
10	1.20	2.17	2.17
25	0.66	1.96	1.96
50	0.57	2.06	2.06
100	0.93	2.44	2.44
200	1.74	3.41	3.41

Table 1. Results when using 40 and 20 for domain and range, respectively (time in seconds)

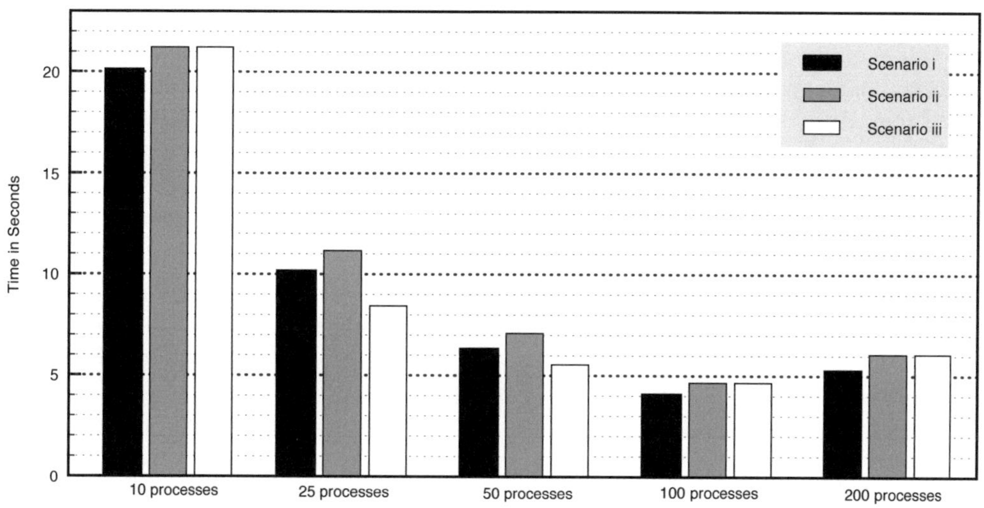

Fig. 6. Evaluating the migration model with AutoMig when using 20 and 10 for domain and range, respectively.

We increase the number of domains when dealing with 20 for the domain's side. This context generated the results presented in Figure 6. As presented in the previous execution, migrations did not take place with 10 processes. They are balanced and their reorganization to the fastest cluster imposed costs larger than the benefits. *pf* of 0.21 was obtained for the current processes-resources mapping by using 20 for domain and 10 processes. All predictions in the

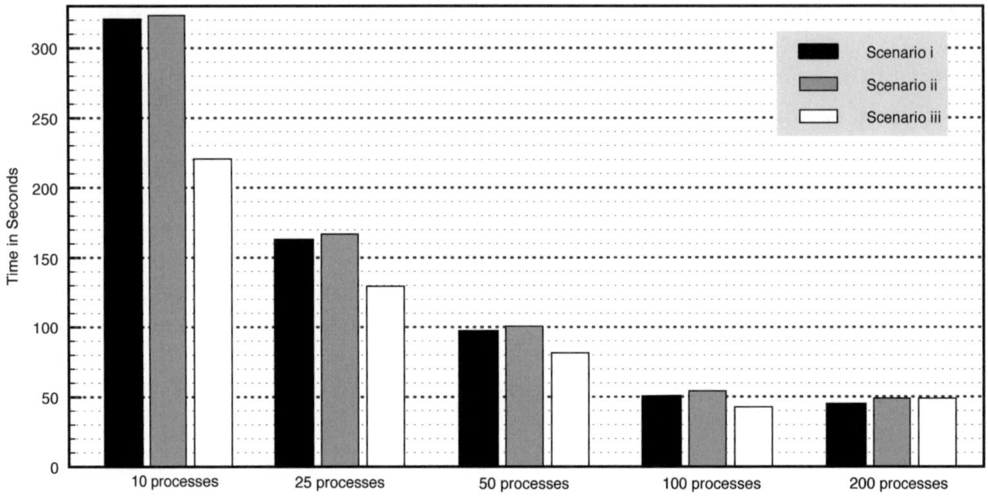

Fig. 7. AutoMig's results when enlarging the work per process at each superstep. This graph illustrates experiments with domain 10 and range 5.

PM list are higher than 0.21 and their average achieves 0.38. However, this configuration of domain triggers migration when using 25 and 50 processes. In the former case, 5 processes from cluster C are moved to the fastest cluster named A. AutoMig's decisions led a gain of 17.15% with process rescheduling in this context. The last mentioned cluster receives all processes from cluster F when dealing with 50 processes. This situation shows up gains of 12.05% with migrations. All processes from cluster C remain on their initial location because the computation grain decreases with 50 processes. Although 14 nodes in the fastest cluster A stay free, AutoMig does not select some processes for execution on them because BSP model presents a synchronization barrier. For example, despite 14 migrations from cluster C to A occur, a group of process in the slower cluster will remain inside it and still limit the superstep's time. Finally, once the work grain decreases when enlarging the processes, the executions with 100 and 200 did not present migrations.

The BSP application demonstrated good performance levels with domain equal to 10 as illustrated in Figure 7. The computation grain increases exponentially with this configuration. This sentence may be viewed through the execution of 10 processes, in which are all migrated to cluster A. Considering that $8((\frac{t}{d})^2.\frac{1}{10})$ express the number of domains assigned to each one of 10 processes, this expression is equal to 500, 2000 and 8000 when testing 40, 20 and 10 values for domain. Using 10 for both domain and the number of processes, the current scheduling produced a *pf* of 1.62. The values of *pf* for the *PM* list may be seen as follows:

- $pf[1..10] = \{1.79, 1.75, 1.78, 1.79, 1.81, 1.76, 1.74, 1.82, 1.78, 1.47\}$.

Considering the first up to the ninth *pf*, we observed that although some processes can run faster in a more appropriate cluster, there are others that remain in a slower cluster. This last group does not allow performance gains due to the BSP modeling. This situation changes when testing the tenth *pf*. It considers the migration of all 10 processes to the fastest cluster and generates a gain around 31.13% when comparing scenarios iii and i. This analysis is illustrated in Figure 8.

The processes from cluster C are moved to A with 25 processes and domain equal to 10. In this case, the 20 other processes stay on cluster L because there are not enough free nodes in the fastest cluster. A possibility is to explore two process in a node of cluster A (each node has 2 GHz) but AutoMig does not consider it because each node in Cluster L has 1.2 GHz. Considering the growth in the number of domains, the migrations with 100 processes becomes viable and get 14.95% of profit. Nevertheless, the initial mapping of 200 processes stands the same position and an overhead of 7.64% in application execution was observed comparing both scenarios i and ii.

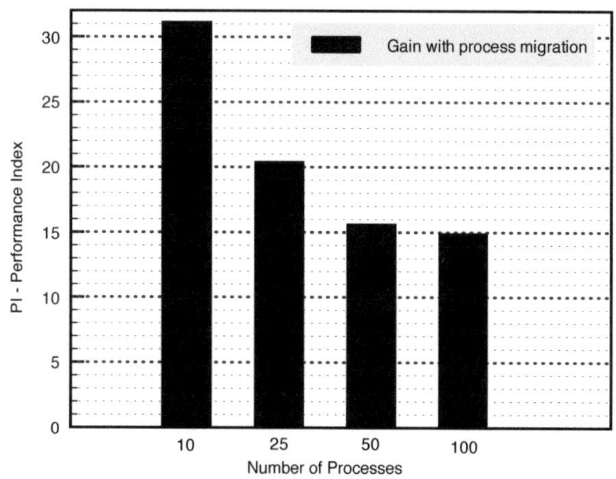

Fig. 8. Migration gains by applying AutoMig with domain 10. Performance Index PI $=(\frac{scen.\ i - scen.\ iii}{scen.\ i} * 100)$.

The previous tests make clear that the higher the computation weight per process, the better will be the gains with process rescheduling. In this way, we tested AutoMig with shorter domain and range values as expressed in Table 2. This table shows the behavior for 10 and 25 processes. Gains about 31.62% and 19.81% were obtained when dealing with AutoMig. In addition, its overhead is shorter than 1%. Observing the results, we can verify that the more time consuming the application, the lower the AutoMig overhead on that. In addition, we verify that the benefits with migrations remain practically constant if we compare the executions with 10 and 4 for the domain values. It is possible to observe that when doubling the number of processes, the application time is not halved as well. There is a limit where the inclusion of processes is not profitable due to the larger number of communication actions and the higher time spent on barriers. Concerning the scalability issue, MigBSP (with AutoMig) shows similar behaviors if compared to those obtained by scenario i (Figures 6 and 8).

Processes	Scenario *i*	Old Heuristic		AutoMig	
		Scenario *ii*	Scenario *iii*	Scenario *ii*	Scenario *iii*
10	12500.51	12511.87	9191.72	12523.22	8555.29
25	6250.49	6257.18	5311.54	6265.38	5011.77

Table 2. Results when using 4 and 2 for domain and range

Table 2 also shows a comparative analysis of the two selection heuristics implemented in MigBSP. We named the one that selects one process at each rescheduling call as Old Heuristic. Despite both obtained good levels of performance, AutoMig achieves better migration results than Old Heuristic (approximately 8%). For instance, 5 processes are migrated already in the first attempt for migration when testing 25 processes. In this case, all processes that were running on Cluster C are passed to Cluster A. This reorganization suggested by AutoMig at the beginning of the application provides a shorter time for application conclusion. In the other hand, 5 rescheduling calls are needed to reach the same configuration expressed previously with Old Heuristic. Lastly, AutoMig imposes larger overheads if compared to Old Heuristic (close to 1%). This situation was expected since two multicast communications among the Set Managers are performed by AutoMig in its algorithms.

6. Related work

Vadhiyar and Dongarra presented a migration framework and self adaptivity in GrADS system (Vadhiyar & Dongarra, 2005). The gain with rescheduling is based on the remaining execution time prediction over a new specified resource. Thus, this framework must work with applications in which their parts and durations are known in advance. In addition, the same problem is shown in the following two works. Sanjay and Vadhiyar (Sanjay & Vadhiyar, 2009) present a scheduling algorithm called Box Elimination. It considers a 3-D box of CPU, bandwidth and processors tuples for selecting the resources with minimum available CPU and bandwidth. The second work comprises the Ding's efforts (Ding et al., 2009). He creates the TPBH (Task Partition-Based Heuristic) heuristic, in which operates with both suffrage and minimum completion time approaches. These two mentioned works treat applications in which the problem size is known in advance. Alternatively, AutoMig just uses data collected at runtime and based on that it takes the performance of different scheduling predictions.

Chen et al. (Chen et al., 2008) proposed processes reassignment with reduced cost for grid load rebalancing. The heuristics permit only movements between the machine with the maximum load level and another machine. Furthermore, this work does not consider the communication issue on selection decisions. Liu et al. (Liu et al., 2009) introduced a novel algorithm for resource selection whose the application reports the Execution Satisfaction Degree (ESD) to the scheduling middleware. Then, this last entity tune the environment by adding/replacing/deleting resources in order to satisfy the user's performance requirements. The main weakness of this idea is the fact that users/developers need to define the ESD function by themselves for each new application.

Concerning the BSP scope, Jiang, Tong and Zhao presented resource load balancing based on multi-agents in ServiceBSP model (Jiang et al., 2007). Load balancing is launched when a new task is insert in the system and is based on the load rank of nodes. The selection service sends the new task to the current lightest node. Load value is calculated taking such information: CPU, memory resource, number of current tasks, response time and number of network links. Silva et al. (da Silva e Silva et al., 2010) explained the resource management on the InteGrade grid middleware. They presented a grid as a collection of clusters, where each one runs its own Cluster Managers (CM). Analogous to MigBSP, CM is responsible for getting data from a cluster and to exchange it among other CMs.

Concerning the migration context, we can cite two works that enable this feature on BSP applications. The first one describes the PUBWCL library which aims to take profit of idle cycles from nodes around the Internet (Bonorden et al., 2005). All proposed algorithms just use data about the computation times from each process as well as from the nodes. Other work comprises an extension of PUB library to support migration (Bonorden, 2007). The author proposed both a centralized and a distributed strategies for load balancing. In the first one, all nodes send data about their CPU power and load to a master node. The master verifies the least and the most loaded node and selects one process for migration between them. In distributed approach, every node chooses c other nodes randomly and asks them for their load. One process is migrated if the minimum load of c analyzed nodes is smaller than the load of the node that is performing the test. The drawback of this strategy is that it can create a lot of messages among the nodes. Moreover, both strategies take into consideration neither the communication among the processes, nor the migration costs.

7. Conclusion

Considering that the bulk synchronous style is a common organization on writing successful parallel programs (Bonorden, 2007; De Grande & Boukerche, 2011; Hendrickson, 2009; Hou et al., 2008), AutoMig emerges as an alternative for selecting their processes for running on more suitable resources without interferences from the developers. AutoMig's main contribution appears on its prediction function *pf*. *pf* is applied for the current scheduling as well as for each level of a Potential of Migration-based list. Each element of this list informs a new scheduling through the increment of one process replacement. *pf* considers the load on both the Sets and the network, estimates the slowest processes regarding their computation and communication activities and adds the transferring overhead of the tested process. The key problem to solve may be summarized in maintaining the current processes' location or to choose a level of the list. AutoMig's load balancing scheme uses the global approach, where data from all processes are considered in the calculus (Zaki et al., 1997). Instead to pay a synchronization cost to get the scheduling information, AutoMig takes profit from the BSP superstep concept in which a barrier always occurs after communication actions.

AutoMig and an application were developed using SimGrid Simulator. We implemented a BSP version of the Fractal Image Compression algorithm. Besides its real utility in satellite and mobile video areas (Guo et al., 2009), this application was taken because it works with a parameter called domain which turns the creation of different load situations easier. Since the application is CPU-bound, the shorter the size of domain the higher the application's time and migration profitability. The results proved this, indicating gains up to 17.15% and 31.13% for domains equal 20 and 10. Particularly, the results revealed the main AutoMig's strength on selecting the migratable processes. It can elect the whole set of processes belonging to a slower cluster to run faster in a more appropriate one. But, sometimes a faster cluster has fewer free nodes than the number of candidates. AutoMig demonstrates that migrations do not take place in this situation, owing to the execution rules of a BSP superstep. It has a barrier that always wait for the slowest process (in this case, the process that will remain on the slower cluster).

Finally, future work comprises the use of AutoMig in a HPC middleware for Cloud computing. This middleware will work on self-provisioning the resources for executing parallel applications. Concerning that each application specifies its own SLA (Service Level Agreement) previously, AutoMig appears as the first initiative to reorganize processes-resources shaping on the fly when SLA fails. If the rescheduling does not solve the performance issue, more resources are allocated in a second instance. The final aim is to reduce the costs on both cloud administrator and user levels.

8. References

Baritompa, W., Bulger, D. W. & Wood, G. R. (2007). Generating functions and the performance of backtracking adaptive search, *J. of Global Optimization* 37: 159–175.
URL: *http://portal.acm.org/citation.cfm?id=1196588.1196605*

Bhandarkar, M. A., Brunner, R. & Kale, L. V. (2000). Run-time support for adaptive load balancing, *IPDPS '00: Proceedings of the 15 IPDPS 2000 Workshops on Parallel and Distributed Processing*, Springer-Verlag, London, UK, pp. 1152–1159.

Bonorden, O. (2007). Load balancing in the bulk-synchronous-parallel setting using process migrations., *21th International Parallel and Distributed Processing Symposium (IPDPS 2007)*, IEEE, pp. 1–9.

Bonorden, O., Gehweiler, J. & auf der Heide, F. M. (2005). Load balancing strategies in a web computing environment, *Proceeedings of International Conference on Parallel Processing and Applied Mathematics (PPAM)*, Poznan, Poland, pp. 839–846.

Casanova, H., Legrand, A. & Quinson, M. (2008). Simgrid: A generic framework for large-scale distributed experiments, *Tenth International Conference on Computer Modeling and Simulation (uksim)*, IEEE Computer Society, Los Alamitos, CA, USA, pp. 126–131.

Chen, L., Wang, C.-L. & Lau, F. (2008). Process reassignment with reduced migration cost in grid load rebalancing, *Parallel and Distributed Processing, 2008. IPDPS 2008. IEEE International Symposium on* pp. 1–13.

da Rosa Righi, R., Pilla, L. L., Carissimi, A., Navaux, P. A. & Heiss, H.-U. (2010). Observing the impact of multiple metrics and runtime adaptations on bsp process rescheduling, *Parallel Processing Letters* 20(2): 123–144.

da Silva e Silva, F. J., Kon, F., Goldman, A., Finger, M., de Camargo, R. Y., Filho, F. C. & Costa, F. M. (2010). Application execution management on the integrade opportunistic grid middleware, *J. Parallel Distrib. Comput.* 70(5): 573–583.

De Grande, R. E. & Boukerche, A. (2011). Dynamic balancing of communication and computation load for hla-based simulations on large-scale distributed systems, *J. Parallel Distrib. Comput.* 71: 40–52.
URL: *http://dx.doi.org/10.1016/j.jpdc.2010.04.001*

Delias, P., Doulamis, A., Doulamis, N. & Matsatsinis, N. (2011). Optimizing resource conflicts in workflow management systems, *Knowledge and Data Engineering, IEEE Transactions on* 23(3): 417 –432.

Ding, D., Luo, S. & Gao, Z. (2009). A dual heuristic scheduling strategy based on task partition in grid environments, *CSO '09: Proceedings of the 2009 International Joint Conference on Computational Sciences and Optimization*, IEEE Computer Society, Washington, DC, USA, pp. 63–67.

Duselis, J., Cauich, E., Wang, R. & Scherson, I. (2009). Resource selection and allocation for dynamic adaptive computing in heterogeneous clusters, *Cluster Computing and Workshops, 2009. CLUSTER '09. IEEE International Conference on*, pp. 1 –9.

El Kabbany, G., Wanas, N., Hegazi, N. & Shaheen, S. (2011). A dynamic load balancing framework for real-time applications in message passing systems, *International Journal of Parallel Programming* 39: 143–182. 10.1007/s10766-010-0134-5.
URL: *http://dx.doi.org/10.1007/s10766-010-0134-5*

Fan, K. (2011). A special parallel job shop scheduling problem, *E -Business and E -Government (ICEE), 2011 International Conference on*, pp. 1 –3.

Goldchleger, A., Kon, F., Goldman, A., Finger, M. & Bezerra, G. C. (2004). Integrade object-oriented grid middleware leveraging the idle computing power of desktop machines: Research articles, *Concurr. Comput. : Pract. Exper.* 16(5): 449–459.

Guo, Y., Chen, X., Deng, M., Wang, Z., Lv, W., Xu, C. & Wang, T. (2009). The fractal compression coding in mobile video monitoring system, *CMC '09: Proceedings of the 2009 WRI International Conference on Communications and Mobile Computing*, IEEE Computer Society, Washington, DC, USA, pp. 492–495.

Hendrickson, B. (2009). Computational science: Emerging opportunities and challenges, *Journal of Physics: Conference Series* 180(1): 012013.
URL: *http://stacks.iop.org/1742-6596/180/i=1/a=012013*

Hou, Q., Zhou, K. & Guo, B. (2008). Bsgp: bulk-synchronous gpu programming, *SIGGRAPH '08: ACM SIGGRAPH 2008 papers*, ACM, New York, NY, USA, pp. 1–12.

Huang, C., Zheng, G., Kale, L. & Kumar, S. (2006). Performance evaluation of adaptive mpi, *PPoPP '06: Proceedings of the eleventh ACM SIGPLAN symposium on Principles and practice of parallel programming*, ACM Press, New York, NY, USA, pp. 12–21.

Jiang, Y., Tong, W. & Zhao, W. (2007). Resource load balancing based on multi-agent in servicebsp model, *International Conference on Computational Science (3)*, Vol. 4489 of *Lecture Notes in Computer Science*, Springer, pp. 42–49.

Kwok, Y.-K. & Cheung, L.-S. (2004). A new fuzzy-decision based load balancing system for distributed object computing, *J. Parallel Distrib. Comput.* 64(2): 238–253.

Liu, H., Sørensen, S.-A. & Nazir, A. (2009). On-line automatic resource selection in distributed computing, *IEEE International Conference on Cluster Computing*, IEEE, pp. 1–9.

Min, L., Xiao, L. & Ying, C. (2009). An establishment and management system of production planning and scheduling for large-piece okp enterprises, *Industrial Engineering and Engineering Management, 2009. IE EM '09. 16th International Conference on*, pp. 964 –968.

Moreno-Vozmediano, R. & Alonso-Conde, A. B. (2005). Influence of grid economic factors on scheduling and migration., *High Performance Computing for Computational Science - VECPAR*, Vol. 3402 of *Lecture Notes in Computer Science*, Springer, pp. 274–287.

Nascimento, A. P., Sena, A. C., Boeres, C. & Rebello, V. E. F. (2007). Distributed and dynamic self-scheduling of parallel mpi grid applications: Research articles, *Concurr. Comput.: Pract. Exper.* 19(14): 1955–1974.

Pascual, J. A., Navaridas, J. & Miguel-Alonso, J. (2009). Job scheduling strategies for parallel processing, Springer-Verlag, Berlin, Heidelberg, chapter Effects of Topology-Aware Allocation Policies on Scheduling Performance, pp. 138–156.

Qin, X., Jiang, H., Manzanares, A., Ruan, X. & Yin, S. (2010). Communication-aware load balancing for parallel applications on clusters, *IEEE Trans. Comput.* 59(1): 42–52.

Sanjay, H. A. & Vadhiyar, S. S. (2009). A strategy for scheduling tightly coupled parallel applications on clusters, *Concurr. Comput. : Pract. Exper.* 21(18): 2491–2517.

Silva, R. E., Pezzi, G., Maillard, N. & Diverio, T. (2005). Automatic data-flow graph generation of mpi programs, *SBAC-PAD '05: Proceedings of the 17th International Symposium on Computer Architecture on High Performance Computing*, IEEE Computer Society, Washington, DC, USA, pp. 93–100.

Tanenbaum, A. (2003). *Computer Networks*, 4th edn, Prentice Hall PTR, Upper Saddle River, New Jersey.

Vadhiyar, S. S. & Dongarra, J. J. (2005). Self adaptivity in grid computing: Research articles, *Concurr. Comput. : Pract. Exper.* 17(2-4): 235–257.

Wang, Z., Wang, F., Gao, F., Zhai, Q. & Zhou, D. (2011). An electric energy balancing model in a medium enterprise grid, *Power and Energy Engineering Conference (APPEEC), 2011 Asia-Pacific*, pp. 1 –4.

Xhafa, F. & Abraham, A. (2010). Computational models and heuristic methods for grid scheduling problems, *Future Gener. Comput. Syst.* 26(4): 608–621.

Xing, C. (2008). An adaptive domain pool scheme for fractal image compression, *Education Technology and Training and Geoscience and Remote Sensing* 2: 719–722.

Yang, C.-T. & Chou, K.-Y. (2009). An adaptive job allocation strategy for heterogeneous multiple clusters, *Proceedings of the 2009 Ninth IEEE International Conference on Computer and Information Technology - Volume 02*, CIT '09, IEEE Computer Society, Washington, DC, USA, pp. 209–214.
URL: *http://dx.doi.org/10.1109/CIT.2009.138*

Yao, L. & Zhu, W. (2010). Visual simulation framework of iron and steel production scheduling based on flexsim, *Bio-Inspired Computing: Theories and Applications (BIC-TA), 2010 IEEE Fifth International Conference on*, pp. 54 –58.

Yu, J. & Buyya, R. (2005). A taxonomy of scientific workflow systems for grid computing, *SIGMOD Rec.* 34(3): 44–49.

Zaki, M. J., Li, W. & Parthasarathy, S. (1997). Customized dynamic load balancing for a network of workstations, *J. Parallel Distrib. Comput.* 43(2): 156–162.

Zhu, Z., Chu, F., Sun, L. & Liu, M. (2011). Scheduling with resource allocation and past-sequence-dependent setup times including maintenance, *Networking, Sensing and Control (ICNSC), 2011 IEEE International Conference on*, pp. 383 –387.

2

Online Production Scheduling and Re-Scheduling in Autonomous, Intelligent Distributed Environments

Edgar Chacón[1], Juan Cardillo[1], Rafael Chacón[2] and Germán Darío Zapata[3]

[1]*Universidad de Los Andes, Mérida*
[2]*Janus Sistemas C.A, Mérida*
[3]*Universidad Nacional Sede Medellín, Medellín*
[1,2]*Venezuela*
[3]*Colombia*

1. Introduction

The search for the automation of continuous production systems that are widely distributed, such as: electricity, oil production, water and others, is a complex problem due to the existence of multiple production units, transport and distribution systems, multiple points of delivery that must be coordinated, the presence of constant changes in the demand and restrictions in the production units. Automation includes constant changes in the production goals that must be captured by the system and every production must self-adjust to these new goals. The production re-scheduling is imperatively on line because it implies evaluating the conditions of every unit in order to know the available capacity and how feasible it is to establish physical connections amongst the different units. In order to achieve one production goal and follow the process on line, in the event of a change in demand or a failure in any unit, an adjustment in the different assignment could be reached and comply with the new conditions of the system. In order to determine the scheduling of every unit, it is important to know its available capacity, production costs and how they should work with each other in order to obtain a product. Once the scheduling is obtained, it must be monitored constantly in order to know its progress and detect when the goal cannot be reached, so a new scheduling is done. The organizational structure of the production process must allow to maintain a knowledge in every production unit, where the acquisition of information and its processing may be centralized and distributed, establishing different approaches to determine the scheduling.

Among the different organizational approaches in manufacturing control Giebels et al. (2001); Heragu et al. (2002), there are the hierarchical approach, the heterarchical approach and the holarchical approach. The hierarchical approach is where the decision flow in the organization is vertical. Cooperation between units is a decision of the immediate superior level. The heterarchical approach is where the decisions are taken horizontally and intelligence is distributed conceptually speaking. The holarchical approach is a mix of the previous approaches. It has the reactivity of the heterarchical approach and the coordination of the hierarchical approach and evolves towards the handling of different transformation activities in order to achieve a goal.

Conventional production systems establish a hierarchical architecture, where decision-making activities, also known as functional coordination calls, include in a superior level production planning and operations management. The immediate lower level includes production scheduling and optimization. In the following level, in descendent order, it is found the process supervision, where it is possible to consider production re-scheduling when disturbances are encountered. After these levels, the production execution levels are found, where the basic control and process functions are. The functions of the three superior levels are centralized, while now there is a tendency to have decentralized controllers in the execution level. Zapata (2011)

Hierarchical systems typically have a rigid structure that prevent them to react in an agile way to variations. In hierarchical architectures, different levels cannot take initiatives. Modifying automated structures in order to add, drop or change resources is difficult, because it requires updating every level in order to recognize the state of the whole system. Furthermore, failures occurred in inferior levels spread to superior levels, invalidating in some cases the planning and affecting the operation of other tasks inherent to the automation, making the system vulnerable to disturbances and its autonomy and reactivity to these disturbances are weak. The resulting architecture is therefore very expensive to develop and difficult to maintain Montilva et al. (2001).

On the other hand, heterarchical systems have a good performance towards changes, and can auto-adapt continuously to its environment. Nonetheless, heterarchical control does not provide a predictable and high performance system, especially in heterogeneous complex environments, where resources are scarce and actual decisions have severe repercussions in the future performance. For this reason, heterarchical control is rarely used in industries.

The objective of this paper is to show how it is possible to implement a scheduling and re-scheduling mechanism based on intelligent autonomous units. These units are capable of knowing in every moment its state, which ensures reactivity of the system towards internal changes and cooperation among different units for scheduling and rescheduling activities.

The chapter is organized as follows: An introduction that corresponds to this Section. The second Section has the features of distributed production systems in the decision-making sense. Section 3 describes the holonic approach and the configuration of holarchies in this approach. Section 4 establishes the scheduling and re-scheduling algorithms. In Section 5 a study case associated with electrical generation is shown. Section 6 comprises conclusions and future works.

2. Distributed production systems

Production systems widely distributed make one category, where every production unit is independent in its decision-making regarding its internal control, and its connection with other systems is based on the product supply with a quality and a flow that is set before hand. The allocation of this compromise of quality and flow is the result of an evaluation that determines the scheduling for that period of time. The product quality and the flow that is supplied, assuming that input availability is reliable (in quality and flow), depend only on the internal conditions of the unit. The connection between distinct units is what allows their reconfiguration in order to obtain a product of any kind.

Among the classic examples of continuous production systems, it is possible to mention oil extraction and electricity generation, which are characterized by the complexity in operations as well as in management, because coordination amongst different subsystems (units) and its reconfiguration are linked to its real-time execution, which guarantees the flexibility in the operation.

2.1 Oil production

In an oilfield, coexist different ways of extraction. A direct way extracts oil by pressure difference in the oilfield and an indirect way extracts oil by pumping oil, raising gas artificially and injecting steam. In-distinctively of the type of extraction from the oil inlet to the oil-crude separator, the oilfield is characterized by having a specific quality of oil. When it reaches the surface, oil brings mud and gas. Every well is connected to a manifold, which picks up the mix in order to make the first liquid/gas separation that takes two different directions. In the first one, gas goes to another manifold that picks the gas coming from other separators. From this collection manifold it goes to a compression center. The compression center discharges the gas in a pipe or tank that has as an input compressed gas coming from compression centers (QP, QI), and outputs like depleted gas (QM) for every compression center, gas that is going to be exported or sold (QE) and gas available for wells (QD). In the second direction, oil passes through a second liquid separation manifold where mud and oil are separated. Oil is sent to a tank farm for another mix with other oil qualities for its distillation. See Figure 1.

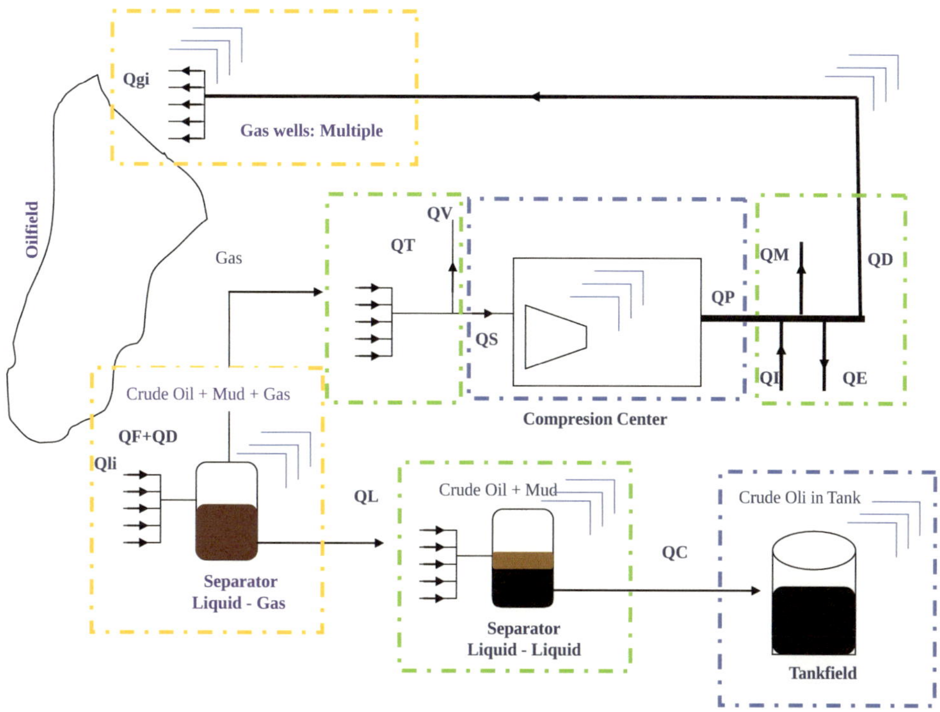

Fig. 1. Oil production example

Where:

QM: Gas depletes
QF: Gas from oilfield
QP: Gas discharging for compressors
QV: Gas sent to the atmosphere
QE: Gas exported or sold
QI: Gas imported from other sources
QD: Gas available in wells
QT: Total Gas
QS: Gas for suction
QL: Oil plus mud
Qli: Oil produced by well *i*
Qgi Gas to well *i*
QC: Total oil produced

This system is considered a complex system due to the number of subsystems and the number of interactions among subsystems. There might be more than one oilfield, a liquid/gas separating manifold may be connected to an oilfield or several oilfields or more than one separating manifold can be connected to one oilfield. In the first direction, a compression center feeds one or more liquid/gas separators. The distribution of compressed gas is done as indicated in the previous paragraph, returning a quality of gas to the wells. In the second direction, when the fluid goes to the oil/mud separation, it is considered that a liquid-liquid separator may be fed by one or more liquid-gas separators. These separators liquid-gas feed one or more mud tanks and one or more tank farms.

2.2 Electricity generation and distribution

In present times, generation of electric current can be done through hydroelectric, thermoelectric, aeolian and nuclear plants and solar panels, among others. Depending on its generating capacity and consumption, these plants must be increasing and decreasing its production in strict real time. Like one source, as theory states, cannot supply all the necessary electric current, these plants must interconnect in a synchronized way in order to supply the consumption needs. In-distinctively of the generating source, the product: electric current, reaches sub-stations, where it amplifies and distributes in high voltage the electric fluid. This fluid in high voltage is transported through distribution networks that transform high voltage in low voltage, towards final users, this is, centers of population, shopping malls, residential complexes, industrial complexes, etc.

The complexity of the operations of this kind of process lies in the generation as well as in the distribution of electric current. Every source of generation is composed of one or more units of generation of the same kind forming a power generation complex. Every generation complex supplies a charge for a determined amount of time and needs to be synchronized in order to be coupled with the system connected with the substation. Substations must distribute to centers of populations as a function of the demanded charge, and this is why the system is considered to be interconnected in a bi-directional way with multiple acceptable paths. See Figure 2.

Fig. 2. Generation and distribution of electricity

In CARDILLO et al. (2009); Chacón et al. (2008), production processes are described, as a network of production units, that we can call nodes. A network can be made of 3 types of nodes, which are: storage units (raw materials, work in process and final products), transportation nodes and production nodes (transformation), as is shown in Figure 3. This network represents the development of the recipe.

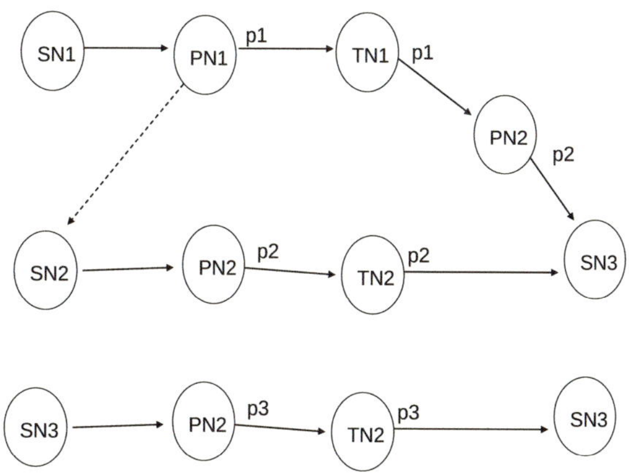

Fig. 3. Nodes for a recipe

The manufacturing of a product is subject to the recipe, which generates a model for fabricating the product, and a model of the product. This recipe specifies the different stages (nodes) where raw materials pass through until it becomes a product. This product model is translated over the physical process of manufacturing, which leads to a configuration of the model in order to follow the product flow according to the recipe. See Figure 4 . This network represents the physical model and its interconnections.

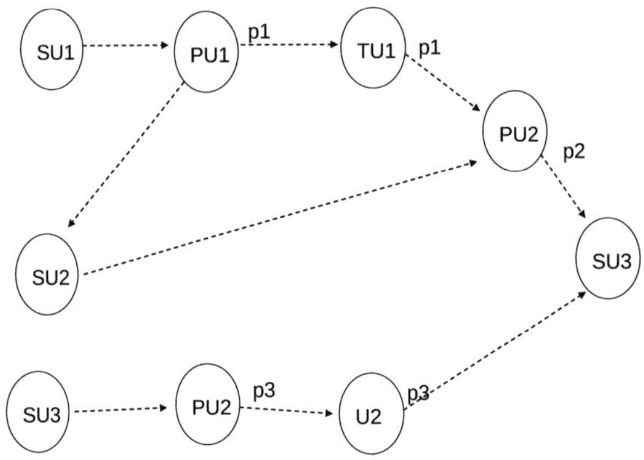

Fig. 4. Nodes representing a physical plant

It is possible (desirable), that when the product flow is placed over the physical process it will generate more than one acceptable configuration, and from those, one is selected in accordance to a criterion and that configuration is established as a pattern in the manufacturing of the product. Because it is a dynamic evolution, at some point there could be a new event (internal or external) capable of altering the performance of the selected configuration and for this reason a re-scheduling in the plan must be done in order to correct the effects of the occurred event. The event can affect, either the execution (regulation, control), the supervision/coordination/management of the device or unit, or a combination of those. The admissibility of a new configuration is established from the set of admissible configurations that have as an initial state the state obtained from the appearance of the event trying to accomplish its mission objective (availability, capacity, interconnection, etc.) of carrying out the recipe or else, putting the process in a secure condition. See figure 5

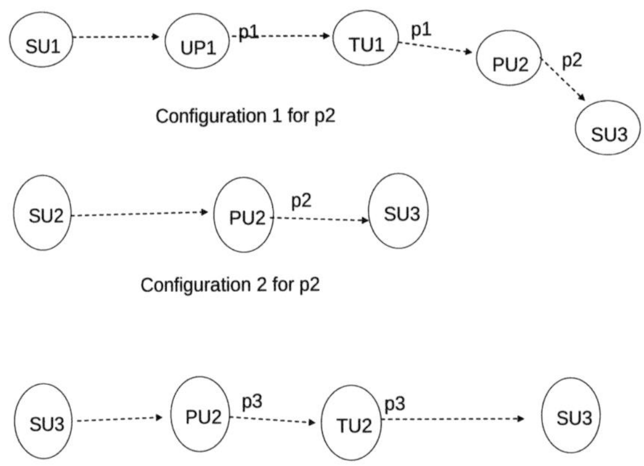

Fig. 5. Nodes representing a production order: Plant Configuration

Even if it is true that until now only a structure that models the product flow and projects itself over the physical process has been mentioned, it is important to describe how these configurations are obtained from the production units.

The production process management is associated by the knowledge itself, this is, knowing the recipes, inputs and their qualities; and the physical process (transformation, transportation and distribution). In the case of process industries, the Business Model is used to describe the chained stages as the input is acquiring added value. Every one of these stages makes the value chain of the process based on the product flow.

CARDILLO et al. (2009); Chacón et al. (2008; n.d.), explains that if a business model and the value chain are used as a global base model of the production process of the company, a production unit can be associated to every link of the value chain. The use of value chains is the base to develop models of different business processes that are specific to the company [26]. A graphic representation of value chains is shown in Figure 6. The product flow can be defined as the different transformation stages that a resource takes (or a set of them) until the final product is obtained.

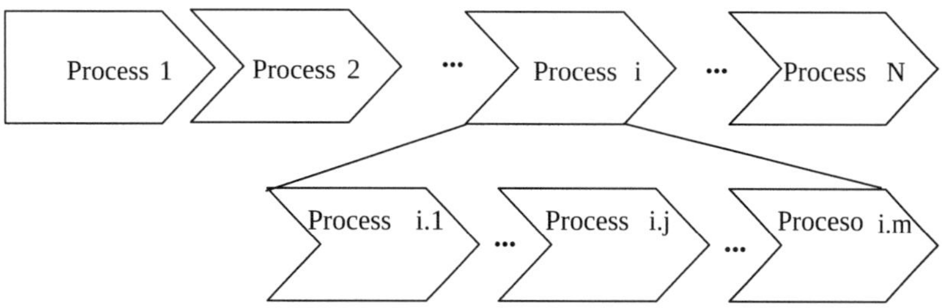

Fig. 6. A value chain in production

2.3 Generic model for Production Units

Every stage or link of the value chain (entry of inputs, processing/transforming and storing), is seen as a Production Unit (PU). So, a productive process or company is an aggregation of cooperating production units. The typification of every production unit depends on how the resource(s) evolves in it, some of them are: Continuous, batch, manufacturing, hybrid, etc. Additionally, every PU performs a specific operation (transformation or transportation) depending on the specifications of the resource (resources) based on the recipe.

Nonetheless, it is possible to find common or generic elements that are common to any PU. Therefore, a PU is distinguished by having:

1. A process for resource uptake (NA)
2. A process for transformation or transportation (NP, NT)
3. A process for storing the transformed product (NA).

Initially, resources and products are managed by the PU coordinator. This is, resources are located and obtained by the coordinator through a PU and the resulting products are shipped towards another PU or the final customer. This way, the process for resource uptake for the PU (NA) is in charge of guaranteeing resources for a given production recipe. After that, the PU selects which should be the production method for making the required transformations for the raw materials. The selection of the configuration and the production method depends on the properties of the resources that enter the PU as well as the resulting products. Once the transformation process is finished, the transformed resource is stored and waiting for another PU to ask for it to be shipped.

In Figure 7 is shown the structural model of a PU, associated with a link in the value chain. The PU manages the production methods, the configuration of the unit and the handling of the resources that intervene on the production process to obtain a product. With this structural description, the bases are placed in order to obtain the necessary information to make the proper negotiations to set a plan. This description allows establishing which are the variables and their respective nominal values, capable of generating the state of the production unit, like: indexes for performance, reliability, product quality, desired quantity of products, production capacity, and storage (maximum and minimum) capacity of the finished product. These values of the state of the PU are key elements to make the negotiations among them, such negotiations are: the production capacity of the PU is given by the capacity of the transformation process, which is determined by the storage capacity, if and only if the raw materials and the rest of the transformation resources are guaranteed.

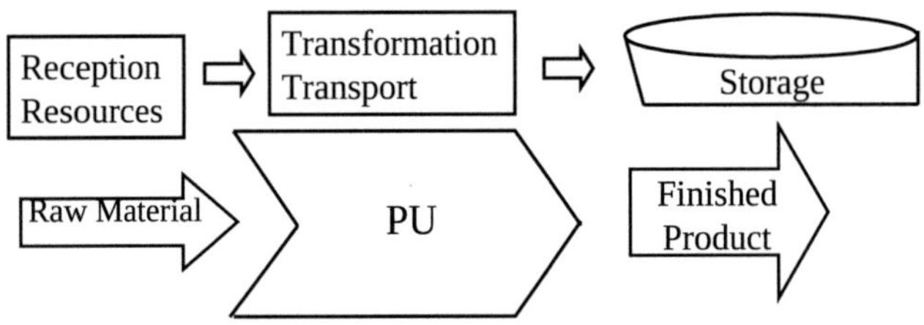

Fig. 7. Model for a Production Unit

The structural model that defines the PU in Figure 8 is made using Unified Modeling Language (UML) Eriksson et al. (2004); Jacobson et al. (n.d.); Muller (1997), where the rectangles represent classes and the lines the relationships among them, using three types of relationships: Generalizations /specializations (arrows), associations (lines) and compositions (arrows with diamonds). In this particular model it is possible to observe the different entities that make and relate with a PU. Particularly, it is shown a special class with the stereotype «association», that is in charge of recording: different resource configurations, productions processes, software for controlling and supervising and its relationship with the production method. It is also shown the classification of resources that are handled by the PU in order to perform the production plan, which supports the planning function. Essentially for space constraints, this structural model is shown in the most concise way; however for every class it is necessary to define attributes, its rules and its operations, in order to support the behavioral model of the PU.

With the description of the given PU, Figure 9 describes the internal built-in proposed model in [30] of the PU. This description follows the rule that a process has control loops. The process set of control loops is called Controlled Process and this is the one that coordinates and manages through the supervisor.

2.4 Process model

It is possible to observe that the flexibility of the production process is associated to the possibility of having a distributed and cooperating process in the sense of the decision-making that makes the PU. This implies understanding the hierarchical scheme and understanding the relationships between the logic process of the recipe and the physical production process including the support areas. This leads to being able to link together the plant model, the product model and the knowledge model, in order to have a process model that must be managed, as is shown in Figure 10.

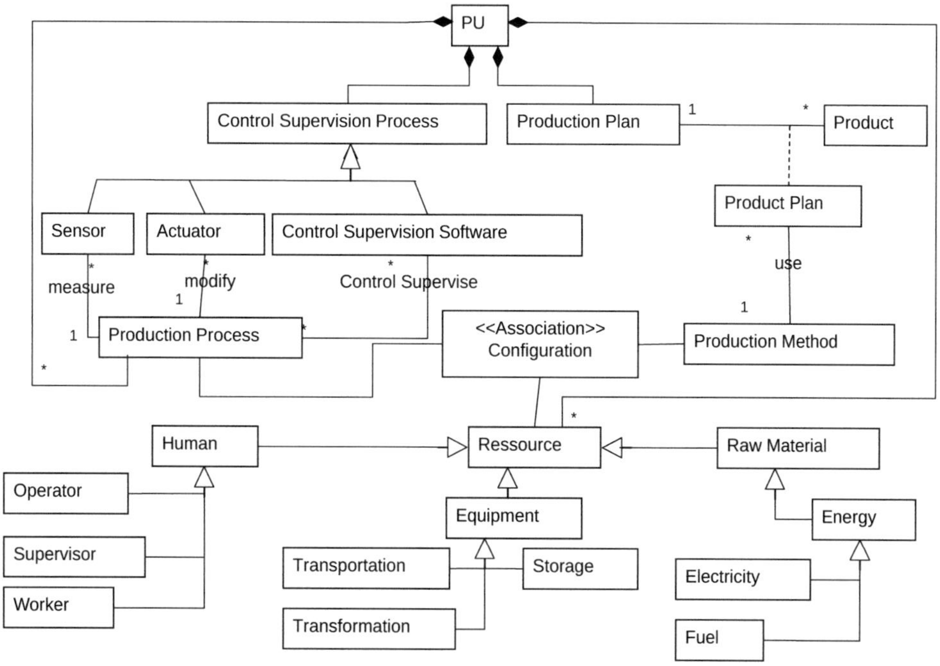

Fig. 8. Structural UML model of a production unit

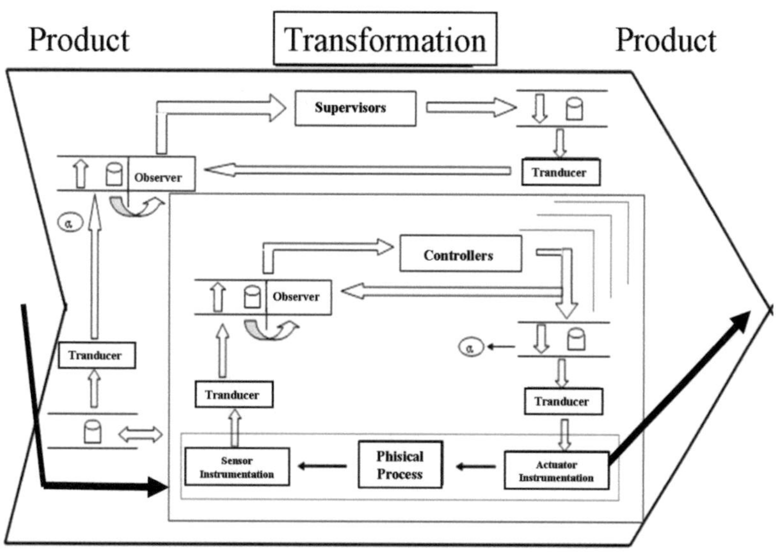

Fig. 9. Built-in model of a production unit

A configuration for the execution of a process (schema model of a process) in Figure 10, can only be obtained if the resources in the physical system, that have a connection among them, have the available capabilities established by the product model during the necessary time for

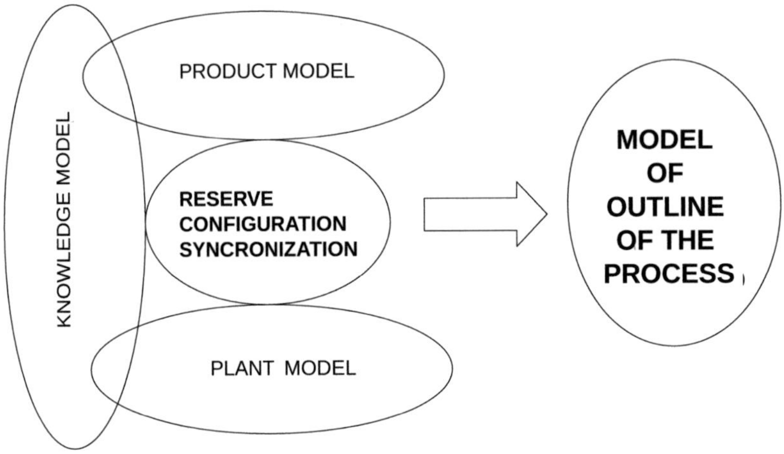

Fig. 10. Procedure to build a configuration

production. The available capability of a resource depends on: a) The need for maintenance and b) tasks assigned to the resource. To the selection of a resource there are other added factors such as reliability and production costs.

3. Holons and holarchy

The current interest lies on the fact that industrial systems need as much high performance as they need reactivity. The answer to this challenge is found in theories about complex adaptive systems. Koestler Koestler (1967) made the observation that complex systems can only succeed if they not only are composed of stable subsystems, every one of them capable of surviving to disturbances, but also are capable of cooperating to form a stable system that is more complex. These concepts have resulted in holonic systems.

The holonic approach starts with the word *Holon* that was introduced by Arthur Koestler Koestler (1967), and was based on the Greek word *holos* that means *whole* and the suffix *on* that means part. According to Koestler, a holon is by itself similar, this is, a structure (Holon) that is stable, coherent and that is composed by structures that are topologically equal to it, Holons, and it, by itself can be a part of a bigger holon and this holon can have many parts that are holons.

Holonic systems combine the advantages of hierarchical systems and heterarchical systems, and at the same time avoid their disadvantages. In order to avoid the rigid architectures of hierarchical systems, holonic systems give autonomy (freedom of decision) to the individual modules (holons). This allows the system to give a quick response to disturbances and the ability to reconfigure itself when it faces new requirements. Additionally, it allows the integration of system modules and a larger range of fabrication systems Zapata (2011).

Compared with holonic control systems, heterarchical control systems may be unpredictable and uncontrollable. This is due to the inexistence of hierarchies in heterarchical systems. It is for this reason that holonic manufacturing systems possess hierarchies, however these hierarchies are flexible, in the sense that they are not direct impositions (hierarchical case), but agreements negotiated by production capacity and response time. This hierarchy differs from traditional hierarchical control in the following:

- Holons can belong to multiple hierarchies,
- Holons can be part of temporal hierarchies, holons do not depend on their own operations but on every holon (associated) in a hierarchy to achieve its goals.

In order to differentiate between strict hierarchies of hierarchical control systems and flexible hierarchies of holonic control systems, the term holarchy has been introduced to identify flexible hierarchies. However, by giving rules and advices, holarchy limits the autonomy of individual holons in order to ensure controllable and predictable performances, unlike heterarchical systems.

The supervision of processes from a holonic perspective possesses properties such as:

Autonomy This property is directly related with decision making. A system will be more autonomous than other while it has more power to take its own decisions. From the supervisory control theory point of view, the autonomy of a holon can be understood as the wider space of states where the supervisor can act.

Reactivity Autonomy ensures that every unit follows the evolution of the process under its control, determining when the goal cannot be reached, internally generating every possible adjustment. Reactivity is satisfied through traceability and control mechanisms that are proper to the system. In case that an internal solution is not possible, the system generates a negotiation scheme.

Proactivity In order to anticipate to situations that put at risk the accomplishment of the mission, holonic supervisory systems have mechanisms that determine in advance the presence of failures, degraded operation conditions or probable breaches when these situations occur. Once these conditions are detected the supervisory system must activate fault tolerance mechanisms that will allow the system to reconfigure itself to respond to this new condition. If the breach is imminent, cooperation mechanisms or negotiation mechanisms are activated

Cooperation From a supervision point of view, cooperation is seen as the physical and logic interaction among components in order to achieve the production goal; this leads every component to have its own goal based on the production goal. This way, cooperation is seen as an enlargement of the space of states among several holons that can interact between them at the instance of the HPU supervisor. If two or more holons can handle between them a disturbance through some pre-established cooperation mechanisms, the intervention of the supervisor in the superior level will not be required, leading to lower interventions from it.

Based on the features presented for the supervisory system of a holonic production system and the features of the conventional approach for process supervision, in Zapata (2011) a comparative chart of the two approaches is presented.

3.1 Structure of a holon in production processes

The structural model of a holon used for production processes, is given by a head, a neck and a body. The **BODY**, is where the transformation/shipping/storage processes are done by a set of physical devices, such as: reactors, compressors, storage units, among others. A **HEAD**, is where the decision making process regarding production is accomplished, based on the knowledge that it has of the production process and the resources. These processes are developed by intelligent systems: Man and machine. The **NECK**, is the interface between the two structures explained before and that is composed by all the informatics structure A UML representation is shown on Figure 13 that stores, supports and transports the information. See Figure 11. A UML representation for a holon is shown on Figure 12.

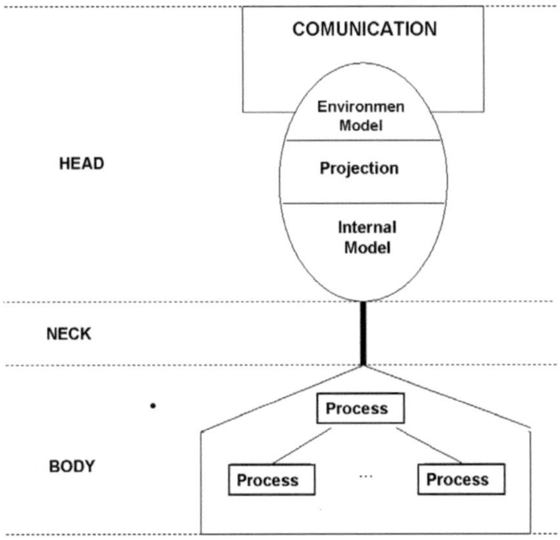

Fig. 11. Structural description of a holon in production processes

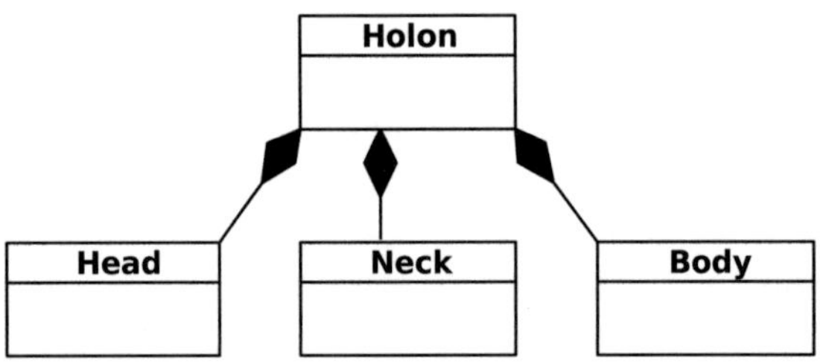

Fig. 12. UML description of a HOLON

As a counterpart, PROSA presents a model of a holon that does not contain a structure associated with the production process, but it contains a description of the elements that make a unit with autonomy be a holon (See Figure 13). These elements are *Order*, *Product* and *Resource*. As it is possible to observe, it does not give information about the interaction between these elements, but it rather helps as a complement to the structural description presented before. These elements are used later on to complete the holonic description presented in Figure 9 that evolves to Figure 11.

One of the most relevant aspects is that until now, several holonic approaches are tied by inheritance to the hierarchical approach Hsieh (2008); McHugh et al. (1995); Van Brussel et al. (1998); Zhang et al. (2003), and this contradicts in a certain way what has been pre-established Zapata (2011), at first, by the holonic description, that for production systems, they must be able to cooperate, allow self-configuration, autonomous, and its complexity seen as an unchanging cooperative embedment of holons, like a distributed system. In Chacón et al. (2008), a proposal for a structure of a Holonic Production Unit (HPU) is proposed, and it

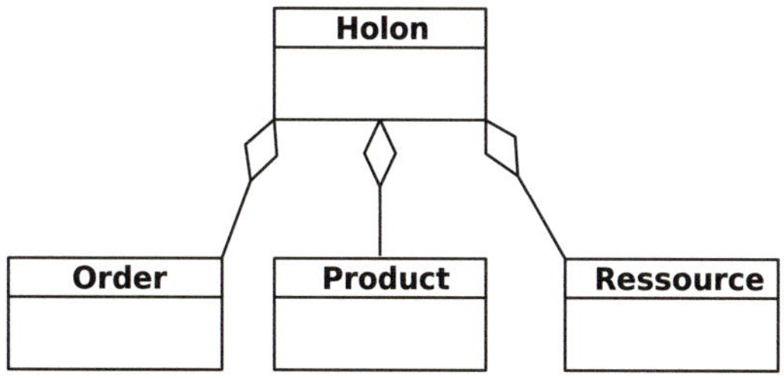

Fig. 13. Holon according to PROSA

is based on the embedded model that is shown in the Figure where it is highlighted the autonomy and the cooperation due to the entry of a negotiator, as it is shown on Figure 14

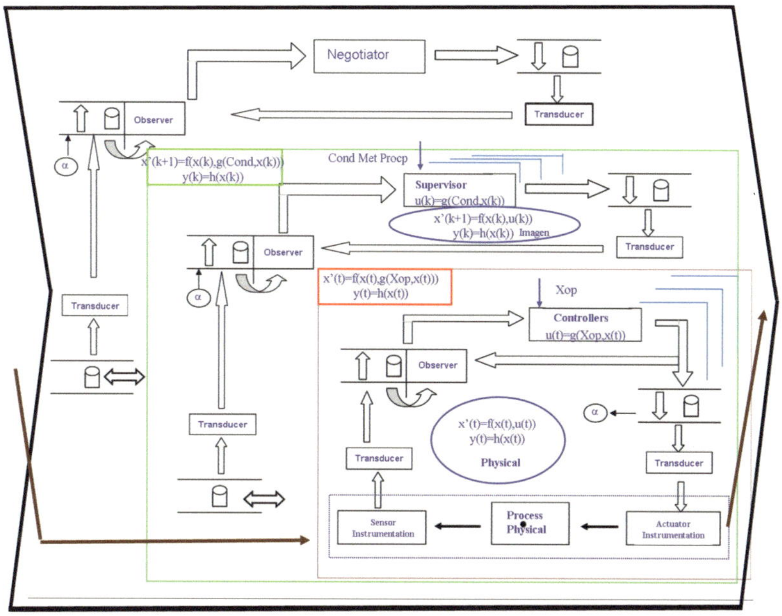

Fig. 14. Embedded model for a Production Unit

In the HPU shown in Figure 14, there is a hierarchical scheme for decision-making that persists Chacón et al. (2008; n.d.). Being able to break the hierarchical scheme to make decisions, goes through understanding the relationships between elements that are part of the productive process (value chain), support areas and the business process. This leads to being able to link the plant model, the product model and the knowledge model, in order to have a process model that is managed, as it is shown in Figure 10.

In order to being able to have a flat model regarding decision-making, the proposed HPU model in 14 is transformed in the model shown in Figure 15.

The autonomy of the Holon Production unit is declared by self-management, this is conceived from the knowledge model of the production unit, therefore, the self-management is nothing more than the planning based on the knowledge of the unit and the one in charge of performing the task is the management system. This leads to a manager agenda. Every element of the agent, leads to the scheduling of each equipment, based on the knowledge of

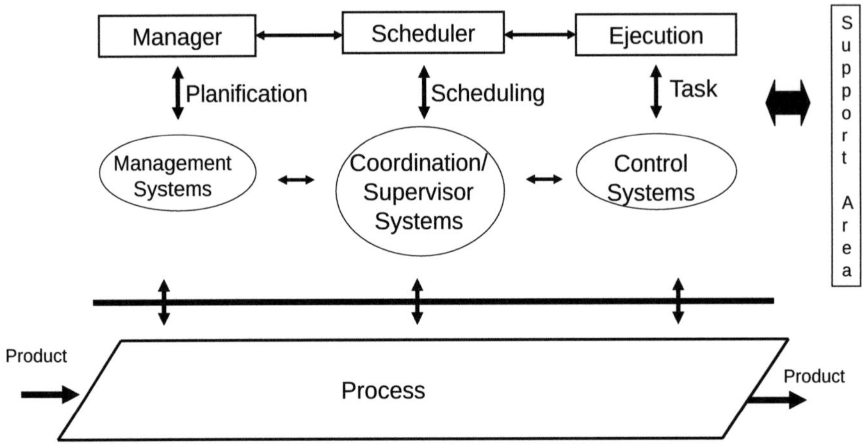

Fig. 15. Holonic model of a Production Unit

every one of them and this is done by the Coordination/Supervision system and its agenda. Every element of the agenda, leads to a set of operations tasks that are based on the knowledge of loops, performed by the control system.

This allows to elucidate that, in general, a *Holon Entity*, in production processes is an entity composed of head, neck and body; it can be planned, scheduled and execute its production goal. This way, the Head is composed by knowledge models of the production unit, agendas and mechanisms for decision making in order to comply with the agenda, therefore with the production goal. The body is in this case the physical process and the neck that is made of the interfaces that communicate both the head and the body. As it was previously said, it is focused in functions such as entity control and product tracking and not in commercial functions that depend in support areas, as it is shown on Figure 15.

3.1.1 The holon equipment and holon unit

The decision components of a holonic system are:

- Planner / Manager that performs planning tasks. It builds different ways to obtain a product, and according to the feasibility of executing activities, it generates an agreement with holons in the superior level
- The scheduler receives the different possibilities and evaluates how feasible they are based on the availability of resources. It generates the supervision scheme that is going to be used by the supervisor in real time
- The Real – Time supervisor, who monitors the progress of the mission, determines, in real time the mission's completion or the possible failures, in order to inform the scheduler and adjust itself if necessary.

With the previous Holon Entity definition for production processes, it is easily shown that it is possible to have devices (Human-machine) capable of handling the knowledge model, this is, descend more intelligence to the equipment level and this way, make them able to plan, schedule and execute its production goal. These are considered the Holon Entity Base, as it is shown in Figure 16.

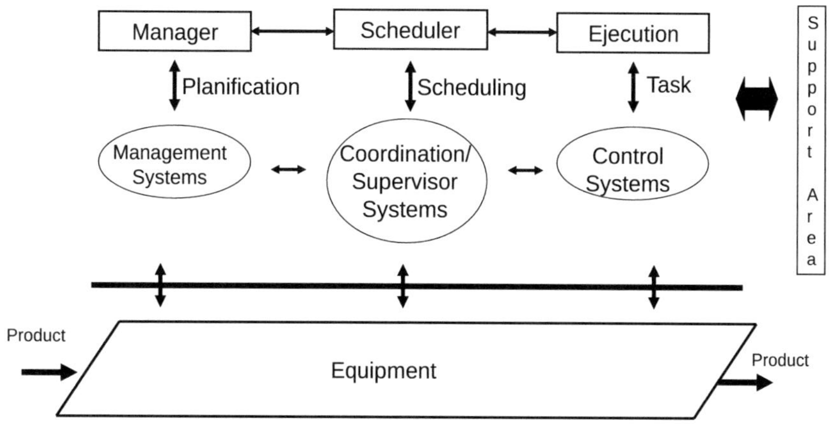

Fig. 16. Holon Equipment

The set of Holons equipment are part of the body Holon Production Unit, as it is shown in Figure 17.

A equipment holon has the necessary intelligence to perform control activities in the shop floor or regulatory control in continuous processes. The control algorithms are sent from the scheduler for the control of the execution. These methods generate events that allow monitoring the production progress and the presence of failures. The local supervisor works as an Event Supervisory System.

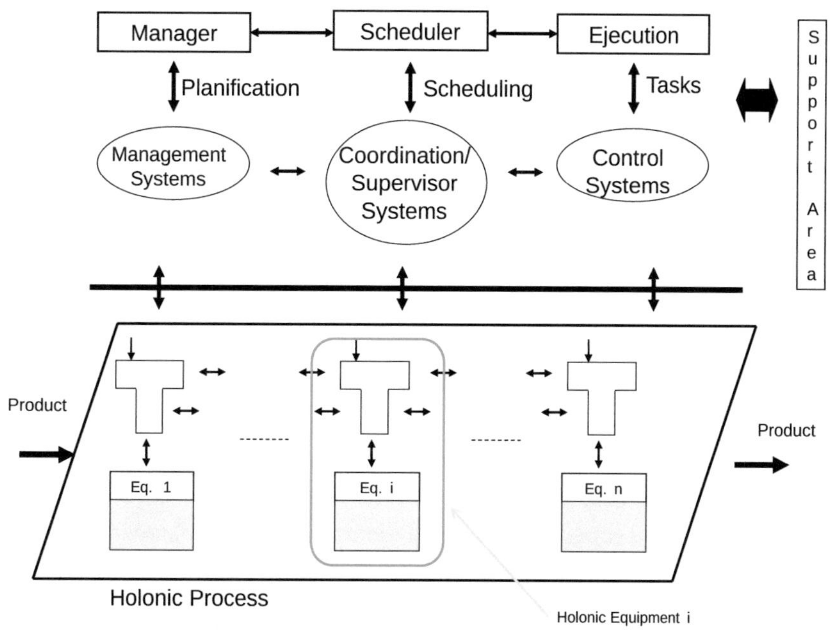

Fig. 17. Holonic Unit

3.1.2 The Holon aggregation

If the same definition of Holon entity is established in production processes, it is possible to establish in a recursive way the Holon Company, the one whose body is composed of Holons Production Unit, as it is shown in Figure 18.

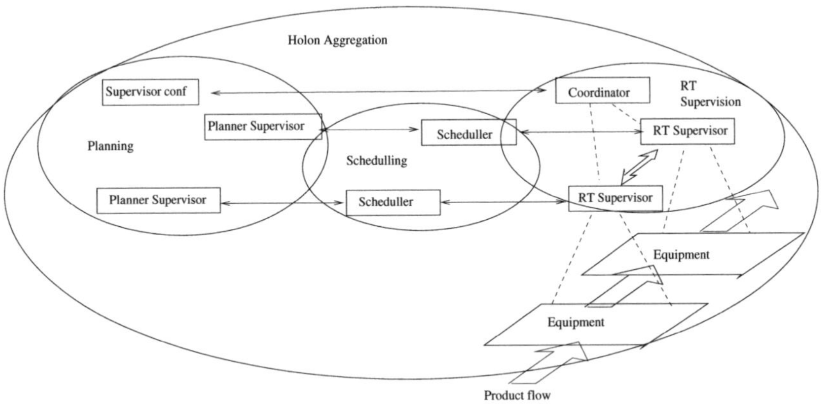

Fig. 18. Holonic Aggregation

Every holon maintains the information of three activities: planning (evaluation of the feasibility to achieve a goal), scheduling (assigning resources and dates for the execution of a task) and following tasks. The product flow is done over the equipment and is supervised internally by every holon (transformation of products and cooperation among the participants holons). A temporary holon that performs coordination tasks between the units and the coordinator, is created. Every holon has a supervisor, who plans and schedules. When the schedule is generated, the structure of the temporary holon that will perform the tracking of tasks is created. Every unit performs physical tasks in production, such as storage and transformation through a set of resources, owned or negotiated in the moment of scheduling, who are the body of a holon.

The logic components for a particular set and its functions are:

- Planner: It selects a method for the set of possible resources and calls for a bid of these resources. It asks the scheduler to calculate the feasibility, and from this information it reserves the selected resources, with the supervision methods that are adequate for the configuration.
- Scheduler: It builds the set of possible configurations, evaluates the configurations and selects the most convenient ones.
- Coordinator: It uses the supervision method determined by the planner and interacts with the body through the supervisors of every equipment or production unit (Holon Resource).

This way, the new holarchy in this holonic conception of Equipment, Units and Company is given by building the body of the holon.

4. Scheduling and re-scheduling algorithms using holons

The scheduling process is associated with the generation of production and work orders according to what was described in Section 3 and the updating mechanism for the necessary information that allows knowing the state of the resources. In Figure 19 it is shown in detail the information flows that are necessary to perform tasks for production scheduling and re-scheduling.

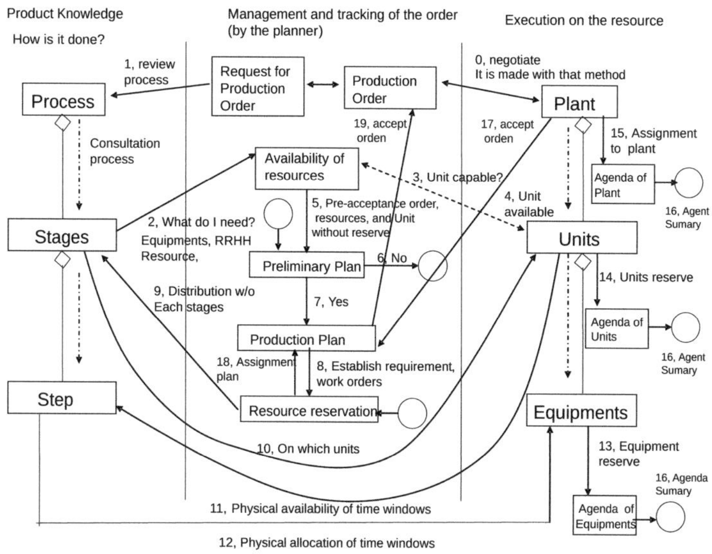

Fig. 19. Handling of a production order

4.1 Scheduling algorithms

The use of holons simplifies the planning process, because every Production Unit knows its available capacity to participate in the production process and the equipments that it has with their physical connections. Every Production Unit has the knowledge to perform a phase of the process and finally obtain the product.

Every PU has the capacity to perform planning, scheduling and tracking functions of the production activities associated with either a phase of the process or its current state. Its state is composed of the agenda, task progress and the state of its own resources, as it is shown in Figure 20

The process to update the agenda (result from planning and scheduling) is as follows: When a production request is received, it is evaluated in order to know if it is feasible to execute. In this process, the Production Unit uses the information from the production method or the product model that is described in terms of a discrete events system Cassandras & Lafortune (2008) (Discrete Events System), similar to the model proposed in ISA 88 and ISA 95 ISA (1995; 2000). It also uses the information from its associated resources (available or not), and their capacities and competencies. The result in this part of the process generates a Petri Net. The reachability tree of this Petri Net is used to verify the feasibility of performing a task. If it is feasible to perform the task, it is incorporated in the agenda. If a resource is a Production Unit, internally it performs the same process. Figure 21

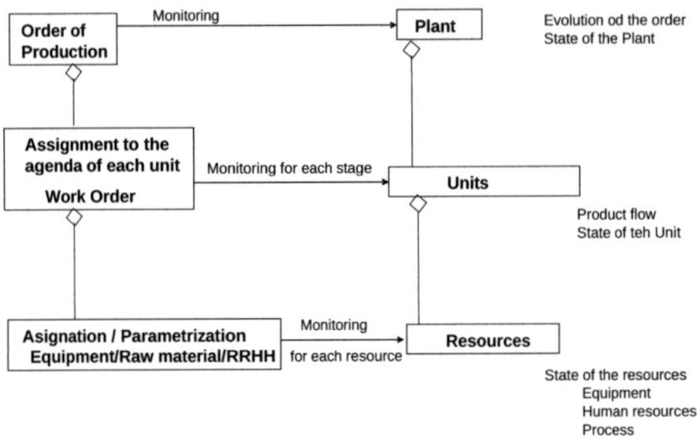

Fig. 20. Concepts associated with the Production Plan

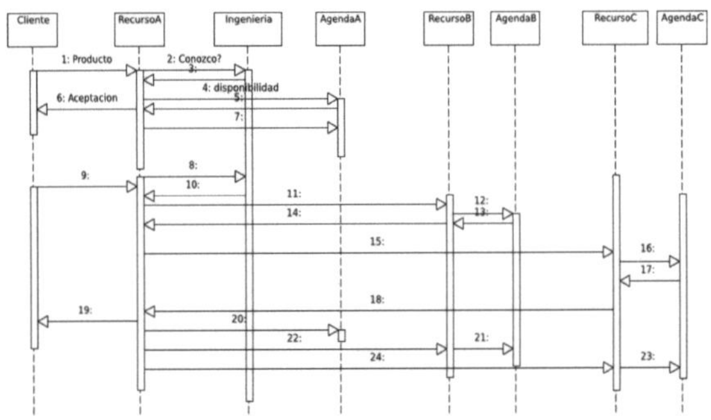

Fig. 21. The negotiation process

Once the order is scheduled, the supervisory system in real time and in terms of a DES Ramadge & Wonham (1989), verifies the fulfillment of the order, generating events that determine its fulfillment, or the need for re-scheduling. See Figure 22.

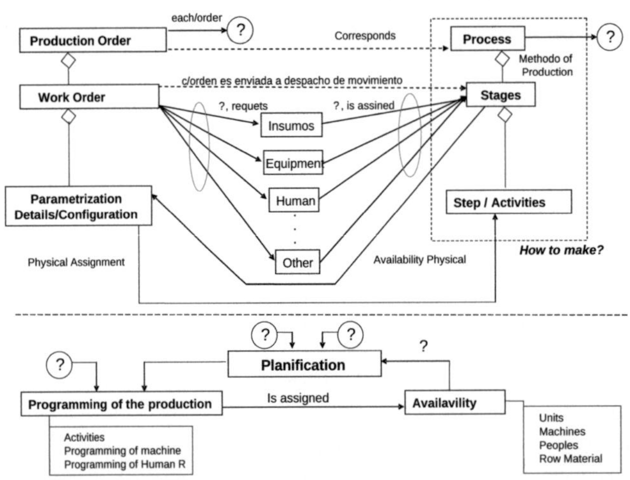

Fig. 22. Following an order

A computational distributed architecture allows determining the set of events that occur in a shop floor, update the discrete state of the tracking system and fire the re-scheduling process, in case any resource presents a failure. If the process cannot be solved within the

PU, it generates a failure event for the PU that contains it and generates a new re-scheduling mechanism from its current state.

5. A case study

The concepts presented about scheduling and re-scheduling in a Holonic Production Unit are illustrated in a thermal power plant with a combined cycle in a 4 x 2 arrangement, this is four gas turbines and two steam turbines. As a complex system, energy generation plants have multiple production units (generation units), transportation systems (gas, steam, oil or water and electrical energy) and present constant changes in the demand and restrictions, physical and operations, among production units. A detailed description of the process is shown in [8]. In Figure 23(a), it is shown a process diagram for a plant with a 2 x 1 arrangement and Figure 23(b), presents a simplified representation of a HPU.

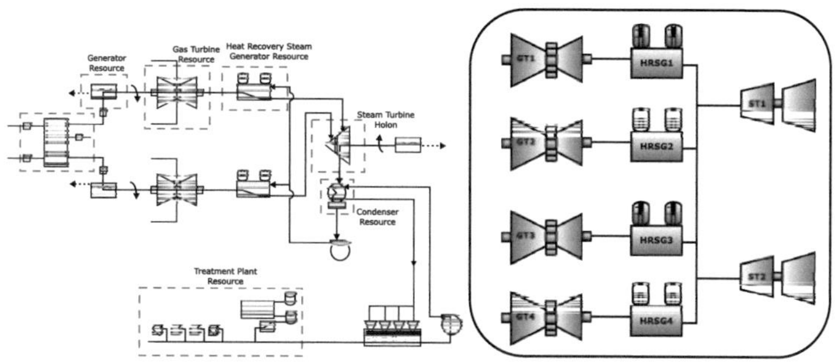

(a) Process diagram 2 x 1 (b) Seen as a HPU

Fig. 23. Thermal power plant with combined cycle

To be able to establish the holons that are part of the HPU, it is taken into account statements presented in 2.3, where it is stated that every holon can be seen as a HPU and has the autonomy to process a product. Thus, "gas turbine-GT" and "Steam turbine ST" are holons. These holons have the ability to negotiate their goals according to their availability and capacity; and have the autonomy and intelligence to perform the process "generate energy". HPUs as resources for services are: boilers, water treatment plants and sub-stations.

Once a process is advancing in the negotiation to reach its production goal and depending on the possible connection among holons, holarchies are formed. The holarchy concept is fundamental to the scheme that responds to disturbances of the holonic paradigm, where every holarchy, once is formed, becomes a HPU contained in a superior HPU (thermal power plant). The production re-scheduling is performed initially inside the holarchy. In the thermal power plant, the possible holarchies are:

$1\,GT + 1\,ST$; $1\,GT + 2\,ST$; $2\,GT + 1\,ST$; $2\,GT + 2\,ST$; $3\,GT + 1\,ST$; $3\,GT + 2\,ST$.

Figure 24(a) shows holarchy $2\,GT + 1\,ST$ and Figure 24(b) shows holarchy $1\,GT + 2\,ST$

Models for production scheduling and re-scheduling are built from Petri Nets (PN) David & Alla (2005); Moody & Antsaklis (1988); Murata (1989), which allow having a representation of the state and the availability of resources, holons and holarchies. The advantage of using PN is that the representation is dynamic and allows establishing a new operational condition from its actual state, which is reached after a disturbance. The details to obtain models are explained in Zapata (2011) and Zapata et al. (2011). About the product

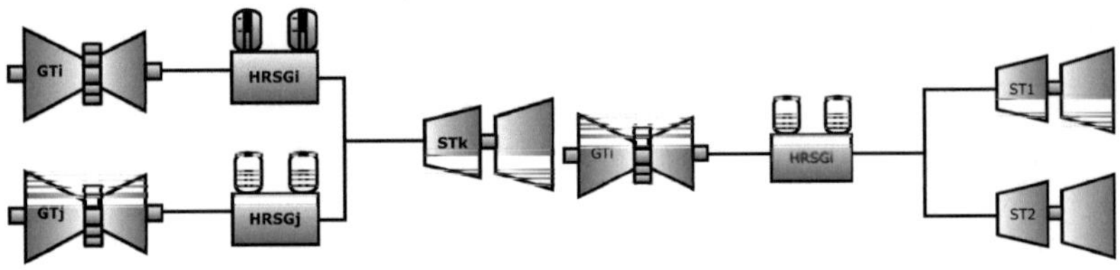

(a) Holarchy 2 $GT + 1\,ST$ (b) Holarchy 1 $GT + 2\,ST$

Fig. 24. Holarchies in a thermal power plant

model, a composition is done to have a PN that represents the global model of the HPU that combines the product, the resource and the connections.

About the product model, a composition is done to have a PN that represents the global model of the HPU that combines the product, the resource and the connections. In Figure 25(a), it is shown a graph that represents the product, with its corresponding PN in Figure 25(b). The resource model is shown in Figure 26.

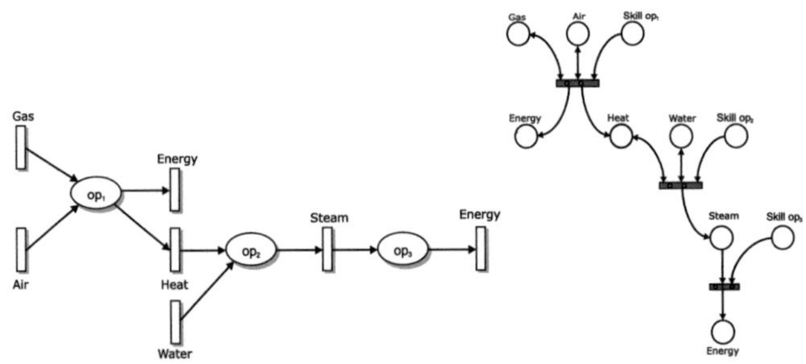

(a) Petri Net for a product (b) Graph of the product

Fig. 25. Modelling the product

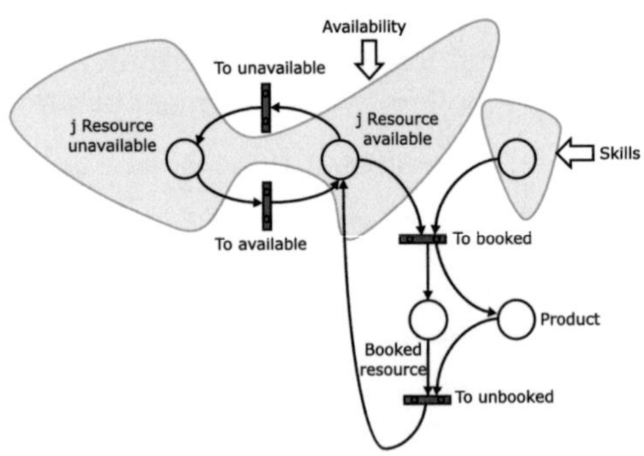

Fig. 26. Resource model

Because of the global model of the HPU, it is shown just a part of the PN model of the holarchy 2 $GT + 1\,ST$ in Figure 27.

The actual state of the resource allows establishing the initial marking to execute the PN, when a negotiation process is launched.

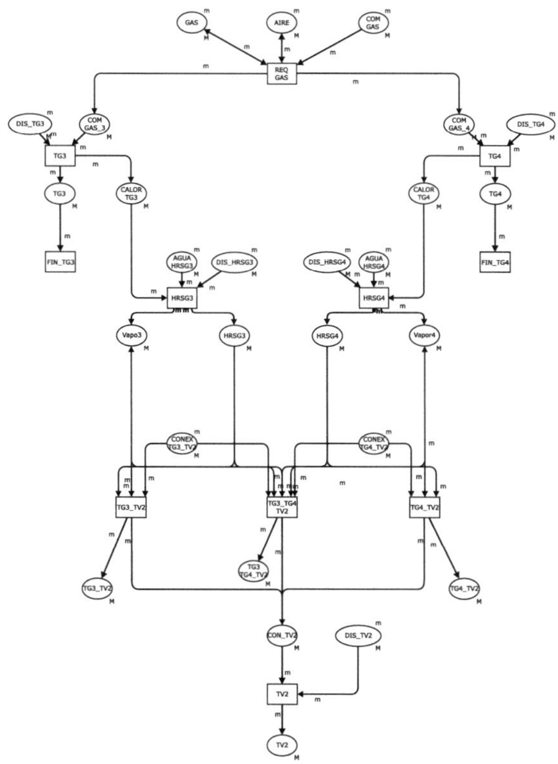

Fig. 27. Petri Net for a Holarchy

The main analysis tool that allows obtaining all of the possible combinations of the resources to reach a goal and determine if it is feasible with the current state of the holons, is the reachability tree Moody & Antsaklis (1988); Zapata (2011). This tree represents a complete state space that a HPU can reach. This space state is of a discrete nature and is obtained through the execution of the PN from its initial marking. Within the holonic conception, the production scheduling method using a reachability tree, allows defining the holarchies that can be grouped in order to perform the mission. In every tree node, capacities of holons and temporary restrictions are added, as it is shown in Figure 28. This tree allows defining the sequence of operations, start and end times, and production plans for every holon and holarchy.

Applying the principles that have been stated, HPUs have a mission, as it is shown in Figure29, where a negotiation between holons and holarchies are done. With the availability of holons, the capacity offered and the costs, as shown in Table 1, mission assignments are presented in Figure 30. Notice that the sum of missions of the different holons is equal to the mission of the HPU. The holarchies formed to perform the mission are given in Table 1:

$H1 = GT1 + GT2 + ST1 \; H2 = GT3 + GT4 + ST2$

If there is a disturbance in the generation period N° 12 and the holon ST2 presents a failure that takes it out definitely from operations, a re-scheduling mechanism based in PN is applied inside the holarchy H2 in order to resolve the disturbance inside itself. The failed mission of the holon is taken totally by Holon GT3.

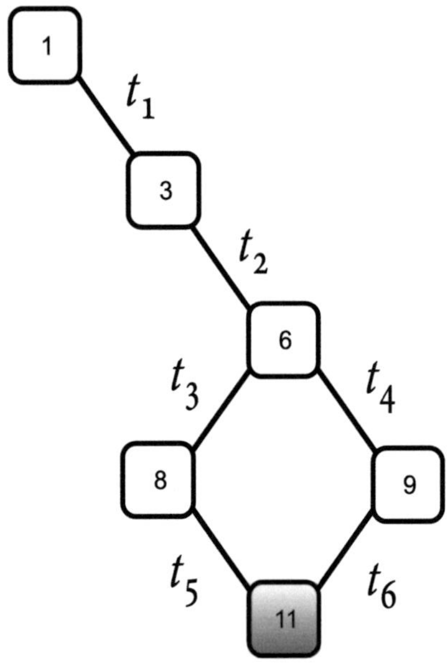

Fig. 28. Reachability tree with capacities and delays

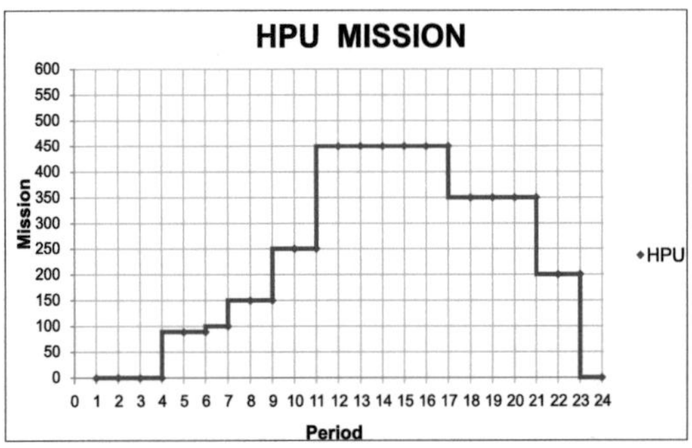

Fig. 29. Mission of the HPU

	Capacity (MW)	Cost (USD/MW
GT1	100	1000
GT2	98	1050
GT3	100	1100
GT4	96	1100
ST1	98	400
ST2	98	450

Table 1. Presentation of offers

The PN model used for production scheduling and re-scheduling evaluates on line the conditions of every production unit and has response times that are appropriate for its application in real time.

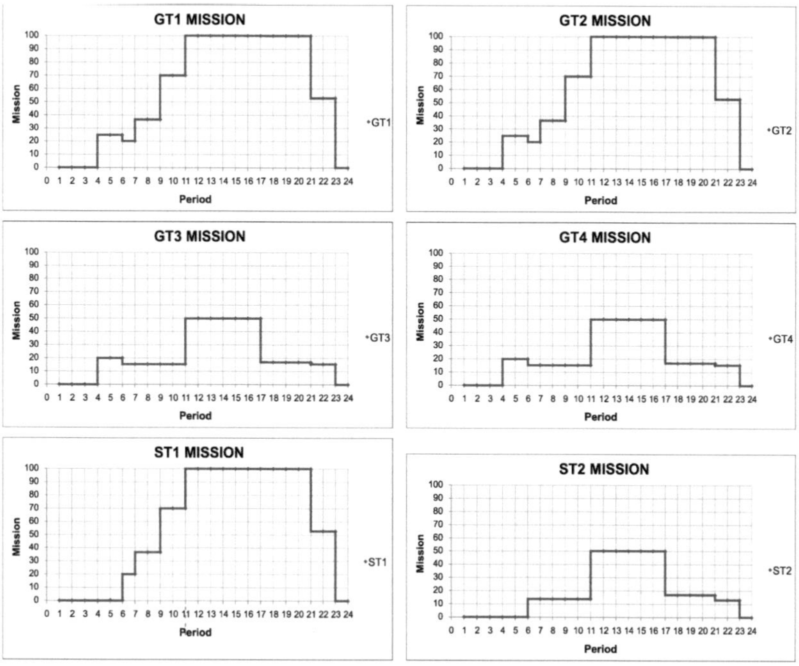

Fig. 30. Mission of the HPU

6. Conclusions

The on line scheduling and re-scheduling of resources in a production system can be achieved in a relatively easy way, if the supervision aspects of the different systems are integrated, as it is proposed by Wohnam-Ramadge, with the scheduling activities. However, it is fundamental that every resource has the autonomy to supervise itself and negotiate with other systems, in order to reach a consensus for the execution of its activities. Every resource must have the knowledge of its capacities, competences and production methods that will allow indicating to the resources that it makes part of, of its available capacity, in the case of scheduling and re-scheduling, and also indicating in an effective and quick way, when it detects that it cannot achieve its goals.

This scheme has been used in the case of manufacturing systems for controlling production shops and in the case of continuous production systems in the generation of electrical energy, as it is shown in the study case.

Now, new schemes are being developed for the description, in an easy way for users, of the physical organization of physical components of the plant and the production units. Software applications are integrated through mechanisms that describe workflows of every unit and the way they cooperate with other units.

7. References

CARDILLO, J., CHACON, E., BESEMBEL, I. & RIVERO, D. (2009). Sistemas holónicos embebidos en procesos de producción continua, *Revista Técnica del Zulia* 32.

Cassandras, C. & Lafortune, S. (2008). *Introduction to Discrete Event Systems*, Springer, New York, NY.

Chacón, E., Besembel, I., Rivero, D. & Cardillo, J. (2008). *Advances in Robotics, Automation and Control*, InTECH, chapter THE HOLONIC PRODUCTION UNIT: AN APPROACH FOR AN ARCHITECTURE OF EMBEDDED PRODUCTION PROCESS.

Chacón, E., Besembel, I., Rivero, D. & Cardillo, J. (n.d.). Holonic production process: A model of complex, precise, and global systems, *Proceedings of ICINCO'07*.

David, R. & Alla, H. (2005). *Discrete, Continuous, and Hybrid Petri Nets*, Springer.

Eriksson, H.-E., Penker, M., Lyons, B. & Fado, D. (2004). *UML 2 Toolkit*, Wiley.

Giebels, M. M. T., Kals, H. J. J. & Zijm, W. H. M. (2001). Building holarchies for concurrent manufacturing planning and control in etoplan, *Computers in Industry* 46(3): 301–314.

Heragu, S. S., Graves, R. J., Kim, B.-I. & Onge, A. S. (2002). Intelligent agent based framework for manufacturing systems control, *IEEE TRANSACTIONS ON SYSTEMS, MAN, AND CYBERNETICS – PART A: SYSTEMS AND HUMANS* 32(5): 560 – 573.

Hsieh, F.-S. (2008). Holarchy formation and optimization in holonic manufacturing systems with contract net, *Automatica* 44(4): 959–970.

ISA (1995). *ANSI/ISA–S95.00.01-1995 Batch Control Part 1: Models and Terminology*, ISA.

ISA (2000). *ANSI/ISA–S95.00.01-2000 Enterprise-Control System Integration Part 1: Models and Terminology*, ISA.

Jacobson, I., Booch, G. & Rumbaugh, J. (n.d.). Specification of the uml, Rational Software. http://www.rational.com/uml/.

Koestler, A. (1967). *The Ghost in the machine*, Arkana Paris.

McHugh, P., Merli, G. & Wheeler, W. A. (1995). *Beyond Business Process Reengineering: Towards the Holonic Enterprise*, John Wiley, New York, N.Y.

Montilva, J., Chacón, E. & Colina, E. (2001). Metas: Un método para la automatización integral en sistemas deÂ producción continua, *Revista Información Tecnológica. Centro de Información Tecnológica* 12(6).

Moody, J. O. & Antsaklis, P. J. (1988). *Supervisory Control of Discrete Event Systems using Petri Nets*, Kluwer Academic Publishers, Boston / Dordrecht / London.

Muller, A. (1997). *Modelado de Objetos con UML*, Eyrolles y Ediciones GestiÃşn 2000, S. A.

Murata, T. (1989). Petri net: Properties, analysis,and aplications, *Proceedings of the IEEE* 77(4): 541 – 580.

Ramadge, P. & Wonham, W. (1989). The control of discrete event systems, *Proceedings of the IEEE* 77: 81–98.

Van Brussel, H., Wyns, J., Valckenaers, P., Bongaerts, L. & Peeters, P. (1998). Reference architecture for holonic manufacturing systems: Prosa, *Computers in Industry – Elsevier* 37: 255 – 274.

Zapata, G. (2011). *Propuesta para la Planificación, Programación, supervisión y Control de la Producción en Procesos Continuos Desde la Teoría del Control Supervisorio y el Enfoque Holónico*, PhD thesis, Facultad de Ingeniería, Universidad de Los Andes.

Zapata, G., Chacón & Palacio, J. (2011). *Advances in Petri net theory and applications*, InTech, chapter Intelligent production systems reconfiguration by means of Petri nets and the supervisory control theory.

Zhang, J., Gao, L., Felix, T. S. C. & Li, P. (2003). A holonic architecture of the concurrent integrated process planning system, *Journal of Materials Processing Technology* 139(1 – 3): 267 – 272.

Part 2

Scheduling Flow Shops

3

Minimizing Makespan in Flow Shop Scheduling Using a Network Approach

Amin Sahraeian
Department of Industrial Engineering,
Payame Noor University, Asaluyeh,
Iran

1. Introduction

Production systems can be divided into three main categories, job shop, flow shop and fixed site. Cellular technology and flexible manufacturing system are the subsystems of job shop. Production scheduling for fixed site are categorized under the title of project planning, so we can conclude that the production scheduling in general are three forms: a) job shop production scheduling b) flow shop production scheduling and c) project production scheduling. In the traditional flow shop scheduling problem, it is assumed that there is only one machine at each stage to execute passing jobs. With the development of hardware, software, and theory in parallel computing, the traditional model of flow shop scheduling is becoming somewhat unrealistic. Defined to capture the essence of parallel computing is the so-called hybrid flow shop model, in which each job has to go through multiple stages with parallel machines instead of a single machine (Havill & Mao, 2002). Scheduling is one of the most important decisions in production control systems. Every production system should have a kind of production scheduling, no matter whether it is managed and organized traditionally or have a systematic and scientific approach to the planning in the production system. If a scientific approach to production planning is organized, we can be sure that a better usage of the resources especially the machinery and the manpower are considered and a better situation for competition are formed in the market. In this chapter we try to use a mathematical optimization model for doing this job. The systems which we are concerned with are two subsystems of the flow shop, which are simple flow shop and hybrid flow shop. The goal is to minimize the total completion time of all the activities and the approach which are used is a linear programming which dominates the heuristic models which mostly used for the NP-Hard problems. to do this, first of all we convert the production system into the network form, then we find the critical activities which affects the total completion time (makespan), then we assign some budget to the activities to crash them, by assigning some budget to some of the operations, (Hojjati & Sahraeyan, 2009) the operation time of these activities reduces and affects the total completion time of all the operations and because of the shortage of the budget, the problem is solved and determines which activities are better to absorb the limited budget to minimize the makespan.

1.1 Definition of production systems

1.1.1 Project production system

A type of non-continuous systems, in which the final goods are completed in a fixed place and the machinery and manpower are moved toward the finished goods. For planning the production in this situation we use a project program like CPM(Critical Path Method), PERT(Program Evaluation and Review Technique) or GERT(Graphical Evaluation & Review Technique). The most characteristic of this system is that, either the movement of the product is impossible (like bridges and roads and ...) or it is very difficult (like aircrafts and large ships).

1.1.2 Job shop production system

A job shop is a type of manufacturing process structure where small batches of a variety of custom products are made. In the job shop process flow, most of the products produced require a unique set-up and sequencing of processing steps. Similar equipment or functions are grouped together, such as all drill presses in one area and grinding machines in another in a process layout. The layout is designed to minimize material handling, cost, and work in process inventories. Job shops use general purpose equipment rather than specialty, dedicated product-specific equipment. Digital numerically controlled equipment is often used to give job shops the flexibility to change set-ups on the various machines very quickly. Job shops compete on quality, speed of product delivery, customization, and new product introduction, but are unlikely to compete on price as few scale economies exist.

When an order arrives in the job shop, the part being worked on travels throughout the various areas according to a sequence of operations. Not all jobs will use every machine in the plant. Jobs often travel in a jumbled routing and may return to the same machine for processing several times. This type of layout is also seen in services like department stores or hospitals, where areas are dedicated to one particular product or one type of service.

A job is characterized by its route, its processing requirements, and its priority. In a job shop the mix of products is a key issue in deciding how and when to schedule jobs. Jobs may not be completed based on their arrival pattern in order to minimize costly machine set-ups and change-overs. Work may also be scheduled based on the shortest processing time.

Capacity is difficult to measure in the job shop and depends on lot sizes, the complexity of jobs, the mix of jobs already scheduled, the ability to schedule work well, the number of machines and their condition, the quantity and quality of labor input, and any process improvements.

1.1.3 Group technology and flexible manufacturing systems

Group technology and flexible manufacturing system are the subsystems of job shop. In cellular group technology, in which the layout is cellular, factory will be broken to units called cells. In these units family parts based on its characteristics, such as parts of geometry, size or similar process are formed. And machinery that have the task of processing on the parts to be allocated cells. Machinery mentioned as possible located close to each other and machinery group is formed. Cellular production is the combination of the two job shop and project production systems. In other words, cellular system has advantages of job shop (production flexibility and diversity of parts) and continuous production (high rate of production) together. Cellular system is one of the most important tools to achieve the lean production.

A flexible manufacturing system (FMS) is a manufacturing system in which there is some amount of flexibility that allows the system to react in the case of changes, whether predicted or unpredicted. This flexibility is generally considered to fall into two categories, which both contain numerous subcategories. The first category, machine flexibility, covers the system's ability to be changed to produce new product types, and ability to change the order of operations executed on a part. The second category is called routing flexibility, which consists of the ability to use multiple machines to perform the same operation on a part, as well as the system's ability to absorb large-scale changes, such as in volume, capacity, or capability.

Most flexible manufacturing systems consist of three main characteristics. The work machines which are often automated CNC(Computer Numerically Controlled) machines are connected by a material handling system which is called AGV (Automated Guided Vehicle) to optimize parts flow and the central control computer which controls material movements and machine flow.

The main advantage of an FMS is its high flexibility in managing manufacturing resources like time and effort in order to manufacture a new product. The best application of an FMS is found in the production of small sets of products like those from a mass production. The other advantages are Faster, Lower- cost/unit, greater labor productivity, greater machine efficiency, improved quality, increased system reliability, reduced parts inventories, adaptability to CAD/CAM(Computer-aided Design/Computer-aided Manufacturing) operations, shorter lead times.

1.1.4 Flow shop production system

Flow shop production system in turn is divided to three main categories: a) simple flow shop, b) hybrid flow shop and c) parallel flow shop. The simple and hybrid flow shop are studied in this chapter and the methodology for parallel flow shop is suggested for further research.

1.1.4.1 Simple flow shop

Much research works both in academic and practical fields have studied flow shop scheduling. In a flow shop system all jobs are processed on machines in the same sequence. However, the processing time of each operation might vary. All jobs are assumed to be ready to be processed at time zero. It is further assumed that there is sufficient physical buffer space between two successive machines without being concerned about the busy or idle status of that machine. A general objective is to develop a schedule that minimizes the makespan. The general flow shop scheduling problem is a NP-Complete problem and a non polynomial time algorithm is expected for these type of problems (French, 1982). The development of heuristic algorithms guarantees good solutions, (Campbel et al., 1970) especially for large size problems. In simple flow shop system there are a set of m machines (processors) and a set of n jobs. Each job comprises a set of m operations which must be done on different machines. All jobs have the same processing operation order when passing through the machines. There are no precedence constraints among operations of different jobs. Usually some assumptions are considered when the theoretical approach is considered. For example operations cannot be interrupted and each machine can process only one operation at a time. These assumptions are considered more deeply in the future. The problem is to find the job sequences on the machines which minimize the makespan, (Khodadadi, 2011) i.e. the maximum of the completion times of all operations (Seda, 2007). As the objective function, mean flow time, completion time variance (Gowrishankar, 2001) and total tardiness can also be used (Pan et al., 2002). The flow shop scheduling problem is usually solved by approximation or heuristic methods. Successful heuristic methods include approaches based on simulated annealing, tabu search, and genetic algorithms (Al-Dulaimi & A.Ali, 2008).

1.1.4.2 Hybrid flow shop

Hybrid flow shop scheduling problems are quite common, especially in the process industry where multiple servers (machines) are available at each stage (Brah & Hunsucker, 1991) as well as in certain flexible manufacturing environments (Zijm & Nelissen, 1990). It is an extension of two classical scheduling problems, the classical flow shop and identical parallel-machine problems. Further, when processing times at a given stage dominate those at other stages, it is natural to increase the system capacity by adding another machine at this stage. A Hybrid Flow Shop (HFS) consists of a series of production stages. Each stage has several machines operating in parallel. Some stages may have only one machine, but at least one stage must have multiple machines. The flow of jobs through the shop is unidirectional. Each job is processed by one machine in each stage and it must go through one or more stage. Machines in each stage can be identical, uniform or unrelated (Linn & Zhang, 1999). The hybrid flow shop scheduling problem is NP-hard and it is usually solved by heuristic methods, that is based on simulated annealing, (Wang et al.) tabu search, and genetic algorithms. There has been a significant amount of research done on the HFS scheduling problem since its first attempt in 1971.

The main difference between the products of this production system and the other two is that the products of flow shop cannot be disassembled. In the other words, we do not have any assembly activity in flow shop systems but in job shop and Project production the product is divided to different components. We can see these components in the final product and finished goods but, in flow shop this issue is not obvious. So without visiting the production system, by observing only the final product, we can conclude whether this product is made in the flow shop or not.

2. Methodology

Makespan is one of the most important criteria in every production systems; it is equal to the total completion time of all the activities. Minimizing this criterion caused better usage of the resources specially machinery and manpower. In both simple and hybrid flow shop, the methodology is to convert the flow shop into a network form, then a linear programming model with the objective of minimizing the total completion time of all the activities are constructed. Minimizing total completion time of all the activities is equivalent to minimizing makespan in the production system. The result is that the sequencing and scheduling of all the activities are determined.

In the next step, some budget is assigned to crash the possible activities. It is possible that we can only assign budget to some of the activities to decrease their times. The amount of budget is always limited, so one question arises, and it is: what are the best activities to absorb this limited budget to minimize the total completion time of all the activities. This causes to add some more constraints to the problem and the result is that the critical activities are determined. The output of the problem is that how much budget should be assigned to which activities to get the best result. The best result is minimizing the total completion time of all the activities in the network which is equivalent to minimizing makespan.

3. Minimizing makespan in simple flow shop scheduling using a network approach

3.1 Assumptions

3.1.1 Assumptions regarding the jobs

- The sequencing model consists of a set of n jobs which are simultaneously available (at time zero, static environment).
- Job r has a predetermined operation times on machine m. m= 1,..., M
- Set-up time is independent of sequence and is included in the processing time.
- Jobs are independent of each other.
- One unit of production for each job is considered.

3.1.2 Assumptions regarding the machines

- Each machine in the shop operates independently.
- Machines can be kept idle.
- All machines are continuously available.
- No machine breakdown is allowable.
- Each machine can process only one operation at a time point.

3.1.3 Assumption regarding the process

- Processing time for each job on each machine is deterministic and independent of the order in which jobs are to be processed.
- Transportation times of a job and set-up times are included in the processing time.

4. Problem methodology

4.1 Problem definition

In this chapter we are faced with a solved n/m/F/C max problem. By using heuristics such as (Campbell et al.1970, Nawaz et al. 1983). To reminimize the makespan, processing times of the operations can be reduced by providing additional resources, which are available at a cost. The processing time of some of the operations can be reduced by assigning a cost. We would like to obtain minimum makespan by reducing processing time of some of these operations. The problem is to find these operations with a pre specified budget.

4.2 Converting flow shop scheduling into a network

Since we are interested in reducing makespan by employing additional resources crashing the project we develop a method to convert a flow shop scheduling into the project network.

Let us define the following notations to convert flow shop scheduling problems into a network.

4.3 Nomenclature

The following terminology is used for modeling the problem:

N: number of jobs.
M: number of machines.
J: an activity number.
m: machine number.
r: job number.
J_{rm}: job r on machine m.
i,j: activity from node i to node j.
Tj: starting time of node j.
Di,j: normal duration time of activity from node i to node j.
Df(i,j): minimum crashing time of activity i to j.
$d_{i,j}$: crashed duration time of activity i to j.
$C_{i,j}$: slope of crashing cost for activity i to j.
B: predetermined budget.

4.4 Network requirements

Certainly when flow shop scheduling problem is converted into a network, it is important to prepare network requirement. Thus, let us define network requirement as follows:

Nodes: Each node represents an event of start or finish operation(s) on machines.

Activity: Activities are the operations to be done on specific machine and have duration equal to processing time.

Predecessors: Activity representing the previous operation for the same job constitutes a preceding activity to the operation. Further the activity corresponding to the operation of the job on the same machine which is before this job in the sequence also constitutes preceding activity.

Duration time: "processing time" is the duration of the activity.

Resources: Machines are the resources.

Suppose we have a flow shop system with n jobs and m machines. The data is described in table 1.

Activity	Predecessors	Duration Time	Machine
1,1	---	D_{11}	M_1
.	.	.	.
.	.	.	.
i,j	(i,j-1),(i-1,j)	D_{ij}	M_j
.	.	.	.
.	.	.	.
r,m	(r,m-1),(r-1,m)	D_{rm}	M_m

Table 1. Data for general model

4.5 Linear programming application to find minimum makespan subject to budget limitation

According to objective that is to select operations to be crashed for finding minimum makespan subject to budget limitation, after converting flow shop scheduling problem into the network, now it is possible to crash the network to find minimum makespan.

4.6 Problem formulation

The problem can be formulated as follows:

$$MinZ = T_n - T_1$$
$$ST.$$

$$\sum\sum C_{i,j}(D_{i,j} - d_{i,j}) \le B \tag{1}$$

$$T_j - T_i \ge d_{i,j} \tag{2}$$

$$D_{f(i,j)} \le d_{i,j} \le D_{i,j} \tag{3}$$

$$T_i, T_j, d_{i,j} = \text{integer}$$

Constraint (1) is related to budget limitation, that additional cost for crashing could not be greater than pre specified budget. Constraint (2) states that the start time of event j should be at least equal to start time of *i* and crash duration of activity *i-j*. Constraint (3) is related to the lower and upper bounds on crash duration.

It is obvious that all t_i's and *dij* must be non negative and also integer. However, due to structure of the problem, it can be solved as a linear programming problem.

4.7 Converting the flow shop problems into a network

In next section we have shown how to convert flow shop problems to project network. We can calculate the earliest start, latest start, floats for all the activities using CPM method. Earliest finishing time of the project is equal to makespan for the flow shop problem. After that we can use the linear programming model for crashing the network.

5. Numerical example

Consider that we have a flow shop problem with four machines and five jobs. According to Campbell et al. (1970), the sequence has been obtained as A-B-C-E-D with corresponding processing times as given in table 2.

		MACHINE			
		Cutting (min)	Pressing (min)	Drilling (min)	Welding (min)
JOB	Part A	5	6	8	4
	Part B	7	5	7	3
	Part C	6	3	5	3
	Part D	8	3	2	3
	Part E	7	4	3	4

Table 2. Processing time of a flow shop with 4 machines and 5 jobs.

Conversion of this problem to a CPM network with information of prerequisite is given in table 3 and the corresponding network is shown in figure 1.

Node (i,j)	Activity (J_{rm})	Predecessor	Duration time ($D_{i,j}$)
1,2	J_{A1}	---	5
2,3	J_{B1}	J_{A1}	7
3,4	J_{C1}	J_{B1}	6
4,5	J_{E1}	J_{C1}	7
5,6	J_{D1}	J_{E1}	8
2,7	J_{A2}	J_{A1}	6
8,9	J_{B2}	J_{B1}, J_{A2}	5
10,11	J_{C2}	J_{C1}, J_{B2}	3
12,13	J_{E2}	J_{E1}, J_{C2}	4
14,15	J_{D2}	J_{D1}, J_{E2}	3
7,16	J_{A3}	J_{A2}	8
17,18	J_{B3}	J_{B2}, J_{A3}	7
19,20	J_{C3}	J_{C2}, J_{B3}	5
21,22	J_{E3}	J_{E2}, J_{C3}	3
23,24	J_{D3}	J_{D2}, J_{E3}	2
16,25	J_{A4}	J_{A3}	4
25,26	J_{B4}	J_{B3}, J_{A4}	3
26,27	J_{C4}	J_{C3}, J_{B4}	3
27,28	J_{E4}	J_{E3}, J_{C4}	4
28,29	J_{D4}	J_{D3}, J_{E4}	3

Table 3. The project network data for corresponding example.

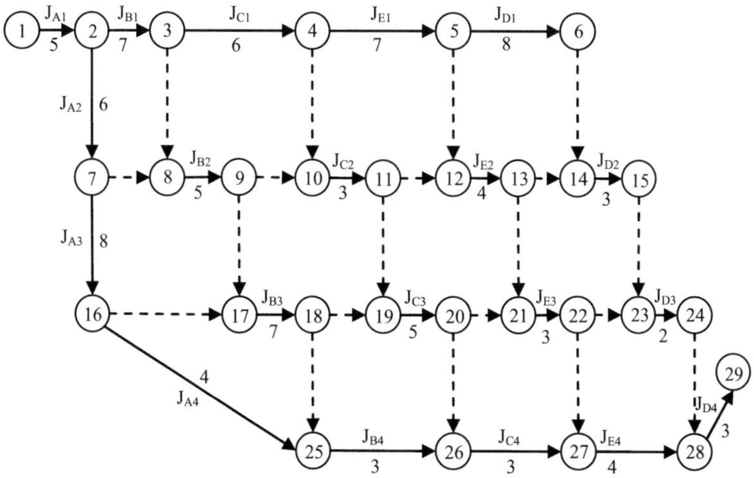

Fig. 1. Project network for given example

- - - -▶ : Dummy operation is used to consider the technological constraints.

Now we should assign some budget to some activities (operation) for which their time can be reduced. These are shown in table 4.

Node (i,j)	Activity (J_{rm})	Predecessor	Duration time ($D_{i,j}$)	Minimum Duration time ($D_{f(i,j)}$)	Cost Slope ($)
1,2	J_{A1}	---	5	4	1150
2,3	J_{B1}	J_{A1}	7	3	1400
3,4	J_{C1}	J_{B1}	6	6	---
4,5	J_{E1}	J_{C1}	7	5	1100
5,6	J_{D1}	J_{E1}	8	8	---
2,7	J_{A2}	J_{A1}	6	4	900
8,9	J_{B2}	J_{B1},J_{A2}	5	5	---
10,11	J_{C2}	J_{C1},J_{B2}	3	3	---
12,13	J_{E2}	J_{E1},J_{C2}	4	4	---
14,15	J_{D2}	J_{D1},J_{E2}	3	3	---
7,16	J_{A3}	J_{A2}	8	6	1000
17,18	J_{B3}	J_{B2},J_{A3}	7	7	---
19,20	J_{C3}	J_{C2},J_{B3}	5	4	1300
21,22	J_{E3}	J_{E2},J_{C3}	3	3	---
23,24	J_{D3}	J_{D2},J_{E3}	2	2	---
16,25	J_{A4}	J_{A3}	4	4	---
25,26	J_{B4}	J_{B3},J_{A4}	3	3	---
26,27	J_{C4}	J_{C3},J_{B4}	3	2	1600
27,28	J_{E4}	J_{E3},J_{C4}	4	4	---
28,29	J_{D4}	J_{D3},J_{E4}	3	3	---

Table 4. Cost of reduced times

5.1 Problem solution

Considering the information given for the problem in tables 3 and 4 and Fig. 1 the objective function and the constraints can be written as follows:

$$MinZ = T_{29} - T_1$$

$$S.\ to:$$

$$1150\times(5-d_{12})+1400\times(7-d_{23})+1100\times(7-d_{45})+900\times(6-d_{27})+1000\times(8-d_{716})+$$
$$1300\times(5-d_{1920})+1600\times(3-d_{2627})\leq 7000$$

$$T_2 - T_1 \geq d_{12} \qquad T_{26} - T_{25} \geq d_{2526}$$
$$T_3 - T_2 \geq d_{23} \qquad T_{27} - T_{26} \geq d_{2627}$$
$$T_4 - T_3 \geq d_{34} \qquad T_{28} - T_{27} \geq d_{2728}$$
$$T_5 - T_4 \geq d_{45} \qquad T_{29} - T_{28} \geq d_{2829}$$
$$T_6 - T_5 \geq d_{56} \qquad T_8 - T_3 \geq 0$$
$$T_7 - T_2 \geq d_{27} \qquad T_{10} - T_4 \geq 0$$
$$T_9 - T_8 \geq d_{89} \qquad T_8 - T_7 \geq 0$$
$$T_{11} - T_{10} \geq d_{1011} \qquad T_{10} - T_9 \geq 0$$
$$T_{13} - T_{12} \geq d_{1213} \qquad T_{12} - T_{11} \geq 0$$
$$T_{15} - T_{14} \geq d_{1415} \qquad T_{12} - T_5 \geq 0$$
$$T_{16} - T_7 \geq d_{716} \qquad T_{14} - T_{13} \geq 0$$
$$T_{18} - T_{17} \geq d_{1718} \qquad T_{14} - T_6 \geq 0$$
$$T_{20} - T_{19} \geq d_{1920} \qquad T_{17} - T_9 \geq 0$$
$$T_{22} - T_{21} \geq d_{2122} \qquad T_{17} - T_{16} \geq 0$$
$$T_{24} - T_{23} \geq d_{2324} \qquad T_{19} - T_{18} \geq 0$$
$$T_{25} - T_{16} \geq d_{1625} \qquad T_{19} - T_{11} \geq 0$$

$$T_{21} - T_{20} \geq 0$$
$$T_{21} - T_{13} \geq 0$$
$$T_{23} - T_{22} \geq 0 \qquad d_{56} = 8$$
$$T_{23} - T_{15} \geq 0 \qquad d_{89} = 5$$
$$T_{25} - T_{18} \geq 0 \qquad d_{1011} = 3$$
$$T_{26} - T_{20} \geq 0 \qquad d_{1213} = 4$$
$$T_{27} - T_{22} \geq 0 \qquad d_{1415} = 3$$
$$T_{28} - T_{24} \geq 0 \qquad d_{1718} = 7$$
$$4 \leq d_{12} \leq 5 \qquad d_{2122} = 3$$
$$3 \leq d_{23} \leq 7 \qquad d_{2324} = 2$$
$$5 \leq d_{45} \leq 7 \qquad d_{1625} = 4$$
$$4 \leq d_{27} \leq 6 \qquad d_{2526} = 3$$
$$6 \leq d_{716} \leq 8 \qquad d_{2728} = 4$$
$$4 \leq d_{1920} \leq 5 \qquad d_{2829} = 3$$
$$2 \leq d_{2627} \leq 3$$
$$d_{34} = 6 \qquad T_i, T_j, d_{i,j} = integer$$

5.2 Results

The results are given using LINGO 7.0 optimally. In table 5, optimum duration for each activity is given. According to budget limitation makespan could be reduced from 41 (before crashing) to 38.

Node (i,j)	Activity (J_{rm})	Crashed Duration time ($d_{i,j}$)
1,2	J_{A1}	4
2,3	J_{B1}	7
3,4	J_{C1}	6
4,5	J_{E1}	5
5,6	J_{D1}	8
2,7	J_{A2}	5
8,9	J_{B2}	5
10,11	J_{C2}	3
12,13	J_{E2}	4
14,15	J_{D2}	3
7,16	J_{A3}	6
17,18	J_{B3}	7
19,20	J_{C3}	5
21,22	J_{E3}	3
23,24	J_{D3}	2
16,25	J_{A4}	4
25,26	J_{B4}	3
26,27	J_{C4}	3
27,28	J_{E4}	4
28,29	J_{D4}	3

Table 5. Optimal duration for each activity according to budget limitation

Table 6 demonstrates different minimum makespan according to different budget assignment.

Budget ($)	**Makespan**
0	41
2000	40
4000	39
6000	38
7000	38

Table 6. Minimum makespan according to the budget

6. Minimizing makespan in hybrid flow shop scheduling using a network approach

6.1 Assumptions

6.1.1 Assumptions regarding the jobs

- The sequencing model consists of a set of n jobs which are simultaneously available (at time zero, static environment).

- Job r has a predetermined operation times on machine m.
- Set-up time is independent of sequence and is included in the processing time.
- Jobs are independent of each other.
- (Shortest Processing Time) SPT rule is used to assign the jobs to the machines.
- One unit of production for each job is considered.

6.1.2 Assumptions regarding the machines

- Each machine in the shop operates independently.
- Machines can be kept idle.
- All machines are continuously available.
- No machine breakdown is allowable.
- Each machine can process only one operation at a time point.
- Each machine starts at its earliest starting time possible.
- Interruption of the machines is not allowed (no repairing during processing).

6.1.3 Assumption regarding the process

- Processing time for each job on each machine is deterministic and independent of the order in which jobs are to be processed.
- Transportation times of a job and set-up times are included in the processing time.

6.2 Nomenclature

The following terminology is used for modeling the problem:

N: number of jobs.
M: number of machines.
J: an activity number.
m: machine number.
r: job number.
s: stage number.
J_{rms}: job r on machine m in stage s.
i,j: activity from node i to node j.
Tj: starting time of node j.
Di,j: normal duration time of activity from node i to node j.
Df(i,j): minimum crashing time of activity i to j.
$d_{i,j}$: crashed duration time of activity i to j.
$C_{i,j}$: slope of crashing cost for activity i to j.
B: predetermined budget.

6.3 Converting H.F.S. into a network model

We can illustrate a general form of H.F.S. with n jobs m machines, and s stages as in Fig. 2

Each operation has a predecessor which is shown in table 7. There are two sets of predecessors, one, the operational constraint, for which every job should be processed in its earlier stage, and second technological constraint for which each machine should operate the jobs in chronological order.

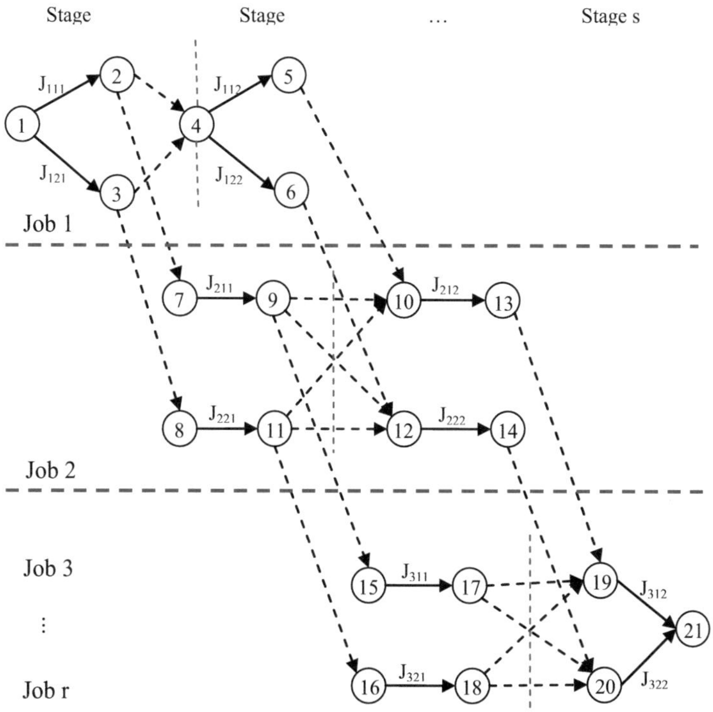

Fig. 2. General model of H.F.S.

Stage	activity	predecessor	Duration time
1	J_{1m1} · J_{rm1} · J_{Nm1}	J_{rm1} m = 1or 2 or … or M , r = 1 J_{rms} m = 1,2,…,M , r = 1 , s = s-1 J_{rm1} m = 1or 2 or … or M , r = 1,2,…,r-1 J_{rms} m = 1,2,…,M , r = r , s = s-1 J_{rm1} m = 1or 2 or … or M , r = 1,2,…,N-1 J_{rms} m = 1,2,…,M , r = N , s = s-1	D_{1m1} · D_{rm1} · D_{Nm1}
2	J_{1m2} · J_{rm2} · J_{Nm2}	J_{rm2} m = 1or 2 or … or M , r = 1 J_{rms} m = 1,2,…,M , r = 1 , s = s-1 J_{rm2} m = 1or 2 or … or M , r = 1,2,…,r-1 J_{rms} m = 1,2,…,M , r = r , s = s-1 J_{rm2} m = 1or 2 or … or M , r = 1,2,…,N-1 J_{rms} m = 1,2,…,M , r = N , s = s-1	D_{1m2} · D_{rm2} · D_{Nm2}
⋮	⋮		⋮

Table 7. Predecessors for general model

6.4 Problem formulation

The problem can be formulated as follows:

$$\begin{aligned}
&MinZ = T_n - T_1 \\
&ST. \\
&\sum\sum C_{i,j}(D_{i,j} - d_{i,j}) \le B \\
&T_j - T_i \ge d_{i,j} \\
&D_{f(i,j)} \le d_{i,j} \le D_{i,j} \\
&T_i, T_j, d_{i,j} = \text{integer}
\end{aligned}$$

7. Numerical example

The methodology is illustrated using a numerical example with 4 jobs, 4 stages and respectively 3,4,3 and 2 machines in each stage. The problem is solved using SPT (Shortest Processing Time). The sequence has been obtained as A-C-B-D with corresponding processing times as given in table 8.

Job	Stage 1			Stage 2				Stage 3			Stage 4	
	M1	M2	M3	M1	M2	M3	M4	M1	M2	M3	M1	M2
Part A	10	12	8	7	10	11	15	9	7	8	20	17
Part B	8	14	11	8	9	14	16	11	6	10	18	19
Part C	13	9	10	13	9	15	8	8	10	6	21	16
Part D	9	11	15	14	16	7	11	9	10	11	19	16

Table 8. Processing time

The network is illustrated in Fig. 3

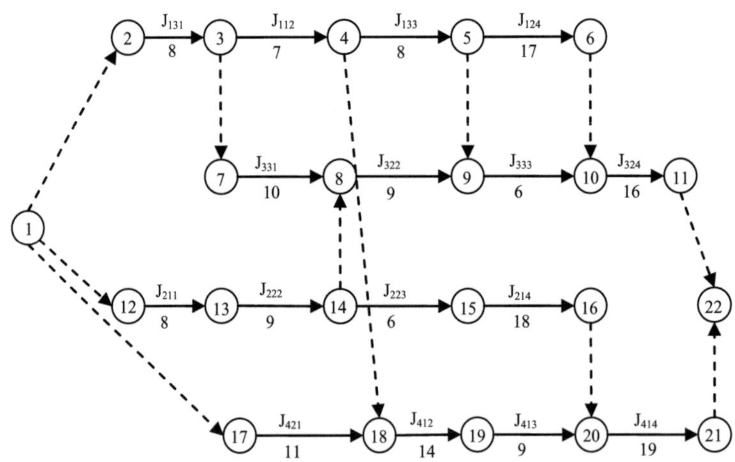

Fig. 3. The network of numerical example

For this problem the predecessors are shown in table 9.

Node (i,j)	Activity (J_{rms})	Predecessor	Duration time ($D_{i,j}$)
2,3	J_{131}	---	8
3,4	J_{112}	J_{131}	7
4,5	J_{133}	J_{112}	8
5,6	J_{124}	J_{133}	17
7,8	J_{331}	J_{131}	10
8,9	J_{322}	J_{331}, J_{222}	9
9,10	J_{333}	J_{322}, J_{133}	6
10,11	J_{324}	J_{333}, J_{124}	16
12,13	J_{211}	---	8
13,14	J_{222}	J_{211}	9
14,15	J_{223}	J_{222}	6
15,16	J_{214}	J_{223}	18
17,18	J_{421}	---	11
18,19	J_{412}	J_{421}, J_{112}	14
19,20	J_{413}	J_{412}	9
20,21	J_{414}	J_{413}, J_{214}	19

Table 9. Predecessor of the numerical example

Now we should assign some budget to some activities (operation) for which their time can be reduced. These are shown in table 10.

Node (i,j)	Activity (J_{rms})	Predecessor	Duration time ($D_{i,j}$)	Minimum Duration Time ($D_{f(i,j)}$)	Cost Slope ($)
2,3	J_{131}	---	8	6	1200
3,4	J_{112}	J_{131}	7	7	---
4,5	J_{133}	J_{112}	8	5	1500
5,6	J_{124}	J_{133}	17	14	---
7,8	J_{331}	J_{131}	10	10	---
8,9	J_{322}	J_{331}, J_{222}	9	9	---
9,10	J_{333}	J_{322}, J_{133}	6	6	---
10,11	J_{324}	J_{333}, J_{124}	16	14	---
12,13	J_{211}	---	8	8	---
13,14	J_{222}	J_{211}	9	7	2100
14,15	J_{223}	J_{222}	6	6	---
15,16	J_{214}	J_{223}	18	15	1800
17,18	J_{421}	---	11	9	---
18,19	J_{412}	J_{421}, J_{112}	14	14	---
19,20	J_{413}	J_{412}	9	7	2000
20,21	J_{414}	J_{413}, J_{214}	19	17	1600

Table 10. Cost of reduced times

7.1 Problem solution

Considering the information given for the problem in tables 9 and 10 and Fig. 3, the objective function and the constraints can be written as follows:

$$MinZ = T_{22} - T_1$$
$$\text{Subject To:}$$

$$1200\times(8-d_{23})+1500\times(8-d_{45})+1800\times(18-d_{1516})+2000\times(9-d_{1920})+2100\times(9-d_{1314})$$
$$+1600\times(19-d_{2021})\leq 10000$$

$$T_3 - T_2 \geq d_{23}$$
$$T_4 - T_3 \geq d_{34}$$
$$T_5 - T_4 \geq d_{45}$$
$$T_6 - T_5 \geq d_{56}$$
$$T_8 - T_7 \geq d_{78}$$
$$T_9 - T_8 \geq d_{89}$$
$$T_{10} - T_9 \geq d_{910}$$
$$T_{11} - T_{10} \geq d_{1011}$$
$$T_{13} - T_{12} \geq d_{1213}$$
$$T_{14} - T_{13} \geq d_{1314}$$
$$T_{15} - T_{14} \geq d_{1415}$$
$$T_{16} - T_{15} \geq d_{1516}$$
$$T_{18} - T_{17} \geq d_{1718}$$
$$T_{19} - T_{18} \geq d_{1819}$$
$$T_{20} - T_{19} \geq d_{1920}$$
$$T_{21} - T_{20} \geq d_{2021}$$
$$T_2 - T_1 \geq 0$$
$$T_{12} - T_1 \geq 0$$
$$T_{17} - T_1 \geq 0$$
$$T_7 - T_3 \geq 0$$
$$T_8 - T_{14} \geq 0$$
$$T_9 - T_5 \geq 0$$
$$T_{10} - T_6 \geq 0$$
$$T_{18} - T_4 \geq 0$$
$$T_{20} - T_{16} \geq 0$$
$$T_{22} - T_{21} \geq 0$$
$$T_{22} - T_{11} \geq 0$$

$$6 \leq d_{23} \leq 8$$
$$5 \leq d_{45} \leq 8$$
$$7 \leq d_{1314} \leq 9$$
$$15 \leq d_{1516} \leq 18$$
$$7 \leq d_{1920} \leq 9$$
$$17 \leq d_{2021} \leq 19$$
$$d_{56} = 17$$
$$d_{1011} = 16$$
$$d_{1718} = 11$$
$$d_{34} = 7$$
$$d_{78} = 10$$
$$d_{89} = 9$$
$$d_{910} = 6$$
$$d_{1213} = 8$$
$$d_{1415} = 6$$
$$d_{1819} = 14$$

$$T_i, T_j, d_{i,j} = \text{integer}$$

7.2 Results

The problem that is formulated in section 7.1 is solved by LINDO software. In table 11, optimum duration (crashed time) and budget used for each activity according to cost slop that is shown in table 10 is given. So with budget limitation of 10000$, we can decrease completion time of production. According to budget limitation makespan could be reduced from 60 (before crashing) to 55.

Activity (d_{ij})	**Crashed Time**	**Budget Used**
d_{23}	7	1200
d_{34}	7	0
d_{45}	8	0
d_{56}	17	0
d_{78}	10	0
d_{89}	9	0
d_{910}	6	0
d_{1011}	16	0
d_{1213}	8	0
d_{1314}	9	0
d_{1415}	6	0
d_{1516}	15	5400
d_{1718}	11	0
d_{1819}	14	0
d_{1920}	9	0
d_{2021}	17	3200
Objective	**55**	**9800**

Table 11. The result of the numerical example

By assigning different budgets, different results can be obtained, this is called the sensitivity analysis of the problem. For example, if we assign 2000$ budget, completion time decrease from 60 to 59, and so. The result can be shown as in table 12.

Budget	**Makespan**
0	60
2000	59
4000	58
5000	57
7000	56
10000	55

Table 12. Sensitivity analysis of the numerical example

8. Conclusions

This chapter reviewed literature on the flow shop scheduling to determine the optimum completion time(minimum makespan). Flow shop system is involved three subsystems, which are simple flow shop, hybrid flow shop and parallel flow shop. In this chapter discussed about only both simple and hybrid flow shop systems. In simple flow shop we use one machine in each stage with identical process for all jobs, but in hybrid flow shop at least in one stage there is more than one machine for processing. According to the literature review, it was found that we have considered the problem of finding minimum makespan for a given sequence of jobs in both simple and hybrid flow shop by using network approach. As a sequence of jobs on machines is known, the problem can be represented as a critical path network. It is shown that both simple and hybrid flow shop problems can be converted to a network model with regard to predecessor relations and processing time.

Then it is estimated the cost of crashing time for each activity, which is possible (cost slope). By using a linear programming formulation the critical activities are determined. Assigning some budget to activities that can be crashed by time, causes to reduce the completion time of all the project or makespan, this by itself causes better use of the resources specially machinery and manpower, which by itself increase productivity.

In addition to above, the important subject in this research is ability to sensitivity analysis. So that, by assigning different budgets, it can be obtained different completion time. Thereupon we can select the optimum completion time considering to corresponding budget.

For further research it is suggested to apply linear programming technique to determine makespan subject to budget limitation in some other systems like parallel flow shop.

9. Acknowledgement

The author is very grateful to Dr. Hojjati who reviewed this chapter and gave some helpful comments.

10. References

Al-Dulaomi, B. & A.Ali, H. (2008). A Novel Genetic Algorithm Approach for Solving Flow Shop Problem. *International Journal of Computer Science and Network Security*, Vol. 8, No.9, pp. 229-235.

Brah, S.A. & Hunsucker, J.H. (1991). Branch and Bound Method for the Flow Shop with Multiple Processors. *European Journal of Operational Research*, No.51, pp. 88-91.

Campbell, H.G.; Dudeck, R.A. & Smith, M.L. (1970). A Heuristic Algorithm for the n Job, m Machine Sequencing Problem. *Management Science*, No.16, pp. 630-637.

French, S. (1982). Sequencing and Scheduling: An Introduction to the Mathematics of the Job Shop. Harwood, Chi Chester.

Gowrishankar, K. (2001). Flow Shop Scheduling Algorithms for Minimizing the Completion Time Variance and the Sum of Squares of Completion Time Deviation from a Common Due Date. *European Journal of Operational Research*, Vol. 132, pp. 643-665.

Havill, J. & Mao, W. (2002). On-line Algorithms for hybrid Flow Shop Scheduling.

Hojjati, S.M.H. & Sahraeyan, A. (2009). Minimizing Makespan Subject to Budget Limitation in Hybrid Flow Shop. *Proceedings of International Conference on Computer and Industrial Engineering 39*, Troyes, France, July 6-8, 2009.

Khodadadi, A. (2011). Solving Constrained Flow-Shop Scheduling Problem with Three Machines. *International Journal of Academic Research*, Vol. 3, No. 1, January, 2011, pp. 38-40.

Linn, R. & Zhang, W. (1999). Hybrid Flow Shop Scheduling: A Survey. *Proceedings of International Conference on Computer and Industrial Engineering 37*, pp. 57-61.

Pan, J.C.H; Chen, J.S. & Chao, C.M. (2002). Minimizing Tardiness in a Two-Machine Flow Shop. *Computer and Operations Research*, Vol. 29, pp. 869-885.

Seda, M. (2007). Mathematical Models of Flow Shop and Job Shop Scheduling Problems. *World Academy of Science, Engineering and Technology*, pp. 122-127.

Wang, H.; Chow, F. & Wu, F. (n.d.). A Simulated Annealing for Hybrid Flow Shop Scheduling with Tasks to Minimize Makespan. *International Journal of Advanced Manufacturing Technology*, Vol.53, No. 5, pp.761-776.

Zijm, W.H.M. & Nelissen, E. (1990). Scheduling a Flexible Machining Centre. *Proceedings of Conference on Engineering Costs and Production Economics 19*, pp. 249-258.

4

Lot Processing in Hybrid Flow Shop Scheduling Problem

Larysa Burtseva[1], Rainier Romero[2], Salvador Ramirez[1], Victor Yaurima[3], Félix F. González-Navarro[1] and Pedro Flores Perez[4]
[1]Autonomous University of Baja California, Mexicali,
[2]Polytechnic University of Baja California, Mexicali,
[3]CESUES Superior Studies Center, San Luis Rio Colorado, Sonora,
[4]Univesity of Sonora, Hermosillo, Sonora,
Mexico

1. Introduction

The scheduling problem in manufacturing systems, where the product components are processed by means of lot units (pallets, containers, boxes) of many identical items, has been recently subject of intense research in modern industry. Most of the common examples in production per lots can be found in serial production like digital devices manufacturing, assembly lines, and information service facilities. The lot processing and the use of batch machines for parallel execution are typical for modern manufacturing systems.

One of the most popular machine environments in scheduling is Hybrid Flow Shop (namely in the following as HFS), in which n jobs must be processed in series of m stages optimizing a given objective function, and at least one stage has parallel machines. In a traditional scheduling problem, a job is indivisible, and it cannot be transferred to the next machine before its processing is finished (Marimuthu et al. 2009; Potts & Baker 1989). On the contrary, the division of jobs into sublots in many situations is permissible and desirable in order to accelerate manufacturing or respect just-in-time (JIT) system through parallel processing. It is known that the HFS problem with $m = 2$ (HFS2) is NP-hard even if one of two stages contains a single machine (Gupta & Tunc, 1998). A HFS with lot processing has additional difficulties such as differences in sublot finish times on machine, unknown lot (sublot) sizes, setup times through lot sequencing, uneven lot sizes under customer request that lead to the demand splitting and allocation problem, etc. These kinds of problems possess a high computational complexity; therefore most of published literature is dedicated to simple flow shop or two-stage HFS machine environment.

Lot processing programming has differences in comparison with traditional job scheduling. On the one hand the job notation depends on the problem assumptions and requires an explicit interpretation; for the other hand one lot as well as one customer's demand can be considered as a job. The theoretical and practical interest on this issue represent models where job splitting is permitted -i.e. one job is allowed to split into sublots to process in parallel on the same or on various machines-. In many models, the sublot size is not clearly evident and must be optimized. When a sublot requires a considerable setup time on machine, the complexity of the problem become increased. The sublots grouping is another important aspect of the lot processing.

The scheduling theory offers two concepts to describe and to solve problems that involve treatment of lots: lot streaming and job batching. These concepts commonly appear together when jobs are allowed to be split. In this chapter, both topics and other relevant problems are discussed in context of HFS machine environment. A solution on a real manufacturing problem is described, where they are involved demand partitioning, lot splitting and batching sublots applied to a HFS2 problem with identical parallel machines at the first stage, and dedicated machines at the second stage.

The following contribution is organized as follows:

- Section 2 introduces the HFS scheduling problem. Some properties and assumptions with certain particularities are discussed and referenced to representative and original scientific contributions in this field of research.
- In Section 3, four major concepts are explained in an in-depth fashion: The Group technology, which embodies jobs or demands according to similar attributes, is aimed to achieve high-volume production efficiency; the batch modeling, where its main definitions are explained and discussed, as long as known scientific papers; and Burn-in operations, a batch processing submodel commonly solicited in semiconductor manufacturing, representing one of the most complex problems in industry; and the Cell architectures that addressed the optimization problem –i.e. minimize production costs and maximize productivity– by the proper arrangement of resources. Additionally, the setup time treatment issue is discussed.
- Lot streaming concept is discussed in Section 4. It is a technic that describes how a job must be split to process and to improve the scheduling result.
- A real manufacturing problem is analyzed in Section 5. A real example in electronic components manufacturing is presented; where the concepts described in this chapter are applied in order to show their usability, in theoretical and practical modeling context.
- Finally, some conclusions in Section 6 are written.

2. Particularities of HFS with lot processing

The HFS scheduling problem is a generalization of the classical flow shop problem by permitting multiple parallel processors in a stage toward to increase the overall capacities or to balance the capacities of the stages, or either to eliminate or to reduce the impact of bottleneck stages on the shop floor capacities (Morita & Shio, 2005). However, the number of variations of this problem is enormous. The HFS differs from the flexible flow line (Kochhar & Morris, 1987) and the flexible flow shop (Santos et al., 1995) problems. In a flexible flow line as well as in a flexible flow shop, available machines in each stage are identical. The HFS does not have this restriction. Some stages may have only one machine, but at least one stage must have a group of machines in parallel. These machines can be identical or generally different. The flow of products is unidirectional. In a classical HFS, each job is processed by at most one machine at each stage and at only one machine at a time. One machine processes no more than one job at each time. The time processing at the stages is known for each job.

The HFS where there are parallel machines at certain stages is very common in industries, which have the same technological route for all products as a sequence of stages. HFSs have important applications in flexible manufacturing systems (FMS), such as electronics and furniture manufacturing, process industries such as chemical, textile, metallurgical, semiconductors, printed circuit board assembly lines, pharmaceutical, oil, food, automobile manufactures, steel making systems, etc. -see, e.g., Tang et al. (2002), Jin et al. (2002), Mathirajan & Sivakumar (2006) and Yaurima et al. (2009)-. Therefore, HFS scheduling has attracted interest by scientific community, being reflected in literature contributions dealing with this particular problem; however the issue of lot processing in this type of shops has not gained sufficient attention.

Let's establish some typical assumptions for a HFS scheduling problem with lot processing:

- There are k stages of processing which occur in a linear order: $1 \ldots k$. Each stage has a predetermined number of parallel machines. However, the number of machines varies from stage to stage.
- A job represents the processing of entire lot of identical pieces, or a demand to produce a several quantity of identical pieces, which are processed in lots.
- Each job visits the stages in unidirectional order. Any stages may be skipped for a particular job but the process flow for all jobs is the same.
- The lot size and the unit processing time on every machine are known in advance and are constant. The sublots belonging to the same lot may have different processing times, a consequence of different sizes.
- Sublots which do not have in-between setup time, form a production batch and are processed together without requiring any machine adjustment. Sublots belonging to different batches may need an essential setup time. A batch machine can process several sublots simultaneously as long as machine capacity is not exceed.
- All sublots contained in the same batch, start and complete at the same time. Interruptions in a sublot and batch processing generally are not allowed.
- Buffers are located between stages to store intermediate products and are mentioned when they are limited.

The three-field notation $\alpha|\beta|\gamma$, named Graham's triplet and recently actualized in (Ruiz & Vazquez-Rodriguez, 2010), is used to describe different HFS variants. The α field denotes the shop configuration, including the shop type and machine environment per stage. The α field is decomposed into four parameter α_1, α_2, α_3, and α_4, and positioned as $\alpha_1\alpha_2$ ($\alpha_3\alpha_4^{(1)}$, $\alpha_3\alpha_4^{(2)}$, ...,$\alpha_3\alpha_4^{(\alpha_2)}$). Here, parameter α_1 indicates the considered shop. The parameter α_2 represents the number of stages in the shop. In this case, a HFS is denoted as FH in the α_1 position and the parameter value of α_2 is major that one. In the notation $\alpha_3\alpha_4^{(k)}$, the parameters α_3 and α_4 describe the type and the number of the machines respectively, of the machine set environments for the stage k. $\alpha_3 \in \{\emptyset, P, Q, R, D\}$, where P indicates identical parallel machines, Q uniform parallel machines, R unrelated parallel machines, and D dedicated parallel machines -see Błażewich et al. (2007) and Pinedo (2008) for in-depth definitions-. The notation D refers to an environment where the jobs to be processed are known in advanced for each machine. A single machine is denoted as $\alpha_3 = \emptyset$. Several authors as Potts and Kovaliov (2000), denotes a flow shop with bath processing as $\tilde{F}$; there are no other specific symbols reported in literature for another shop types.

The parameter β -when used- lists the shop properties, which enumerate specific constraints and assumptions of the problem. The most common model properties associated with a HFS problem with lot processing are listed in Table 1.

Property	Description
Batch (Batch processing)	A machine is able to process up to b jobs continuously without any setup.
$b_k = m_k$	The mk batch processing machines at the stage k.
Fmls (Job families)	The jobs belong to different job families. Jobs from the same family are processed on a machine one after another without any setup in between.
lot	Lot processing.
prmp	The preemptions of jobs are allowed.
R (Removal time)	Machines become free only after the setup of the job has been removed.
split	A job (lot) splits into several parts (sublots) so that their operations may be overlapped. In case of shop with lot processing this parameter refers lot streaming.
s_{si} (Sequence-independent setup times)	The setup time of machine depend only on the job to process.
s_{fg} (Sequence-dependent family setup times)	The setup time of machine to process job (batch) belonged to the family g depends on the previous job (batch) family *f*.
w_j (Weight or importance of job *j*)	That is the priority factor of the jobs in the system.

Table. 1 Model properties associated with a HFS problem.

Finally, the γ field provides the criterion to be minimized. The most solicited Criterion to be minimized in a HFS scheduling problem, is the completion time. This particular objetive function takes place when the last job leaves the system, and is labeled as *makespan* or C_{max}. Another common objetive functions are, among others:

1. $F_{\max}$ as maximum flow time.
2. $L_{\max}$ as maximum lateness.
3. $T_{\max}$ as maximum tardiness.
4. $E_{\max}$ maximum earliness.

Continuing with the context to the task at hand, the triplet $FH2(RM^{(2)}) \,|\, lot^{(1)}, p_j^{(1)} = p, batch^{(1)}, s_{fg}^{(1)}, split^{(2)}, s_{fi}^{(2)} \,|\, C_{\max}$ denotes the problem of scheduling jobs in a two-stage HFS with one batch machine on the first stage and unrelated parallel machines on the second. There exist, on the first stage, a few conditions that must present: lot processing, equal processing times of lots, batching of lots and sequence-dependent family setup time. Lot streaming into sublots and sequence independent setup times are presented on the second stage. The goal is to achieve a *makespan* minimum.

An example of another classification scheme with explicit number of batch processing machines is problem $\tilde{F}2 \mid b_1 = 1; b_2 = 2 \mid C_{max}$, which denotes the *makespan* minimization in a two-machine flow shop, where the first machine is a classical machine, and the second is a batch machine that process up to two jobs simultaneously. Another example is the problem $1 \mid s_{fg} \mid \Sigma C_j$, which denotes the total completion time minimization on a single (classical) machine, where there are job families and sequence-dependent family setup times (Potts & Kovaliov, 2000). It must be noted that due to the novelty of this class of models and scarcity of related work, there exist a lack of conventional notations for the scheduling problems with lot processing.

3. Batch processing

3.1 Group technology

The manufactured products on a plant frequently have technical similarities; therefore can be sorted out into groups according to their design or manufacturing attributes, such as part shape, size, surface texture, material type, raw material estate. The technical similarities of the products within a group permit reduce essentially the number of setups on a machine. Consequently, manufacturing time is decreased and machine usage time is improved.

This idea was adapted as Group Technology (GT). The GT is an approach solicited in manufacturing and engineering management, that in general aims to achieve the efficiency of high-volume production by exploiting similarities of different products and activities in their production/execution (Cheng et al., 2008). The concept of GT is based on the simplification and standardization process, and according to Burbidge (1975) appeared at the beginning of 20th century. Numerous manufacturing companies have taken advantage of GT to improve the productivity and competitiveness -see Wemmerlov & Hyer (1989), Tatikonda & Wemmerlov (1992), Hadjinicola & Kumar (1993) and Gunasekaran et al. (2001)-. The first publications on scheduling in GT environments are trace back to Petrov (1966).

The GT originally emerged as a single machine concept that was created to reduce setup times (Mitrofanov, 1966). Then it was extended to the HFS problem with setup times dependent on the job sequence (Li, 1997). Andrés et al. (2005) introduced the concept of *coefficient of similarity* between each of the products, whose original rol was as parameter, allowing products to be grouped through a heuristic method; and contrary to the basic concept of *exploiting similarities* (taken from the GT philosophy). The first approach allows design engineers to retrieve existing drawings to support the design standardization of new parts and make an accurate cost estimation. The second one produces improvements to the control process, reduction of the setup time and standardized process plans (Kusiak, 1987).

From the GT paradigm, two important concepts arise, product family and batch. The jobs are supposed to be partitioned into F families, $F \geq 1$. A *batch* is a set of jobs of the same family that can be processed jointly (Brucker, 2004). Batching occurs only if setup costs or times are not negligible and several jobs of the same family have to be produced. The processing time depends only on the family of the batch. When the processing is performed in batches of identical items (lots), the processed operations are executed simultaneously. Thus, the completion time of all the jobs in a batch is the finishing time of the last job in the batch. Once the processing a batch is started, it cannot be interrupted, nor can other jobs be added into the batch. The motivation for batching jobs is to gain in efficiency: the processing jobs in a batch maybe cheaper or faster than to individual processing (Potts & Kovaliov, 2000). The term of *family* denotes initial job partitioning, while the term of batch is used to denote a part of the solution. The task to calculate the *batch size* is to decide how many units must be processed consecutively. In Liu & Chang (2000) is indicated that batch sizes must be optimized, because the processing in large batches may increase the machine utilization and reduce the total setup time. However, large batch processing increases the flow time. Therefore, a tradeoff between flow time and machine utilization by selecting batch size and scheduling comes into discussion. According to the GT, no family can be split, only a single batch can be formed for each family.

Many publications use the term batch to denote the initial job partitioning and they use different names like sub-batch, lot, sublot, etc., to denote a set of jobs of the same family processed consecutively on the same machine.

3.2 Batch models

Batch setup models are partitioned into *batch availability* and *job availability* models (Potts & Kovaliov, 2000). According to the batch availability model, all the jobs of the same batch become available for processing and leave the machine together. For example, this situation occurs if the jobs in a batch are placed on a pallet, and the pallet is only moved from the machine when all of these jobs are processed. An alternative assumption is job availability (usually known in the literature as item availability), in which a job becomes available immediately after its processing is completed and completion times are independent of other jobs in the batch.

The processing time of a batch is calculated according to Lushchakova & Strusevich (2010) as follows:

- In Serial batching, also known as *s-batch* or "sum-batch", the processing time of a batch is equal to the total processing times of its jobs.
- In Parallel batching, also known as *p-batch* or "max-batch", the processing time of a batch is equal to the largest processing time of its jobs.

When jobs sizes are considered, the case is usually called *a problem with bounded batches* if the total sizes of the jobs, contained in a batch, must not exceed the capacity of the batch, i.e. $b > n$. As far as each job may have a different size, the number of jobs in each batch may be different. On the other hand, if any number of jobs is allowed to be inserted in a batch, it is an called *unbounded batch*, respectively, $b \leq n$. In this case batches are not restricted in processing any number of jobs (Yazdani & Jolai, 2010).

When one batch is completed, the recourse has to be adjusted for the next batch. Time needed for the setup activities depends on the families of both adjacent batches. A batch is called *feasible* if it can be processed without any tool switches.

3.3 Batch processing machines

In literature, parallel batching scheduling is known as batch processor scheduling or scheduling of *batch processing machine* (BPM). A BPM processes several jobs simultaneously.

The different jobs can be batched together but the processing time of the batch is given by the longest processing time among all jobs in the batch. The BPMs are encountered in many different environments such as chemical processes performed in tanks or kilns and burn-in operations in semiconductor industry. Problems related with scheduling BPMs has been received much attention in scheduling literature in recent years -see Lee et al. (1992), Uzsoy (1994), Li (1997), Brucker et al. (1998), Lee & Uzsoy (1999), Damodaran & Srihari (2004), Mathirajan & Sivakumar (2006), Damodaran et al. (2007) and Manjeshwar et al. (2009)-. Results of these researches are relevant to the HFS, although refer to the single BPM, parallel BPMs, or flow shop BPMs. Uzsoy (1994) described an application for burn-in operations in semiconductor manufacturing. In papers of Damodaran et al. (2007), Liao & Huang (2008), Manjeshwar et al. (2009), it have been provided applications of BPMs in the chemical treatment stage in rim (for bike) manufacturing facilities and in chambers for the environmental stress screening in the printed circuit board assembly environment, respectivelly.

The two important decisions made on BPMs are:

- Grouping part families into batches, and
- Scheduling the batches to improve a performance measure.

The main classification of BPMs is related with *incompatible job families* vs. *compatible job families* (Perez et al., 2005). In the first model, only products belonging to the same family may be processed simultaneously. Uzsoy (1995), Kempf et al. (1998), Dobson and Nambimadon (2001) developed deterministic algorithms to schedule BPMs with incompatible job families. In the second model, it is assumed that products belonging to alternative families may be processed simultaneously -see Lee et al. (1992) where it is modeled burn-in oven as BPM-. Due to the complexity of BPM problems, scheduling research almost focus on single and parallel BPMs (Perez et al., 2005).

Quard and Kuhn (2007) investigate a s-batch scheduling problem for a HFS. Each job belongs to a specific product type. Setup costs are incurred when changing a machine of one product. On each stage, all jobs have the same process time. The objective is to minimize setup costs and the mean flow time. A target number of setups -i.e. parallel machines- for each product type and product type sequence are calculated for scheduling all production stages. The main focus of this paper is to derive the analogy of the scheduling problem to a two-dimensional packing problem and the development of a solution procedure that uses this analogy to solve the original HFS scheduling problem. Genetic algorithms are used as a framework to incorporate these ideas.

Xuan and Tang (2007) addressed the s-stage HFS problem of scheduling n jobs with s-batch processing at the last stage, and reduced to a two-stage HFS. The objective is to minimize a given criterion with respect to the completion time. When the jobs are grouped at the stage s, each batch l has a given size b_l -i.e. consists of b_l jobs-. The batch size can be different for all batches. All the jobs from the same batch must be processed on a machine at stage s consecutively while satisfying given precedence constraints among the jobs within this batch. Each job j has a weight and the waiting of job processing between two adjacent stages causes a penalty cost. A sequence-independent setup time is considered separate from the processing time before the first job of batch l starts processing. It could be anticipatory, meaning that the setup of the next batch can start as soon as a machine becomes free to process the batch. Transportation times are also considered separate from the processing time. This paper establishes an integer programming model and proposes a batch decoupling based Lagrangian relaxation algorithm for this problem solution.

A two-stage HFS scheduling problem in a metal-working company is studied by Luo et al. (2009). The first stage consists of multiple parallel bounded-batch machines with job availability model, and the second stage has only one machine. The setup time is separated from job processing time and depends upon preceding job. A blocking environment exists between two stages with no intermediate buffer storage. Preventive maintenance and machine breakdown are presented. Two types of machine unavailability namely deterministic and stochastic case are identified in this problem. The former occurs on a stage two machine with the start time and the end time known in advance. The latter occurs on one of the parallel BPM in stage one and a real-time rescheduling will be triggered. Minimizing the makespan is considered as the objective to develop the optimal scheduling algorithm. A genetic algorithm is used to obtain a near- optimal solution. The computational results with actual data are favorable and superior over the results from existing manual schedules.

A two-stage HFS with several identical bounded p-batch processing machines is considered in the paper of Bellanger & Oulamara (2009). The problem is motivated by the scheduling of tire in the manufacturing industry. A compatibility relation is defined between each pair of tasks, so that an undirected compatibility graph is obtained which turns out to be an interval graph. The goal is to make batching and sequencing decisions in order to minimize the makespan. Since the problem is NP-hard, several heuristics are developed along with their worst cases analysis. The case in which tasks have the same processing time on the first stage also is considered, and a polynomial time approximation scheme (PTAS) algorithm is presented.

The review described in the above lines, shows that scientific literature in this addressed problem has small number of contributions. One of the main factors that could lead to this condition is that the BPM entails a high computational complexity. Then only one or two stages shops are investigated.

3.4 Setup time treatment

The structure of the breakdown time when a job belongs to a machine includes three phase as follows (Cheng et al., 2000):

1. Sequence independent/dependent setup time that is independent/dependent on the job to be processed.
2. Processing time of the job.
3. Removal time that is independent/dependent on the job that had just been processed.

The setup time is defined as the time required to shifting from one job to another on a given machine. There are separable and non-separable from the process operation. The *non-separable* setup times are either included in the processing times or are negligible, and hence are ignored. There exist some situations in which the non-separable setup and removal operations must be modeled and closely coordinated. Such situations are common in automatic production systems which involve intermediate material handling devices, like automatic guided vehicles and robots, loading and unloading (Crama, 1997; Kim et al., 1997). The *separable* setup times are not part of processing operation.

When separable setup/removal times are not negligible in the scheduling problem, they should be explicitly treated. In many real-life industrial problems such as surface mount technology or printed circuit board manufacturing, job setup is not part of processing time and the required time is sequence-dependent. Cheng et al. (2000) presents an interesting review of flow shop scheduling research with setup times.

The separable setup times could be *anticipatory* (*detached*) or *non-anticipatory* (*attached*). A setup is anticipatory if it can be started before the corresponding job or batch becomes available on the machine. In such a situation, the idle time of a machine can be used to complete the setup of a job on a specific machine. Otherwise, a setup is non-anticipatory, and the setup operations start only when the job arrives at a machine as long as the setup is attached to the job. Furthermore, setup time of a job at a specific machine could be dependent on the job immediately preceding that job or be independent of it.

The setup may reflect the need to change a tool or to clean the machine. As in a family scheduling model, the jobs are partitioned into families according to their similarity, so that no setup is required for a job if it belongs to the same family of the previously processed job. However, a setup time is required at the start of the schedule and on each occasion when the machine switches from processing jobs in one family to jobs in another family. In such model, a batch is a maximal set of jobs that are scheduled contiguously on a machine and share a setup.

The next setup analysis is proposed in the paper of Potts & Kovalyov (2000). Let $\{1, \dots, n\}$ denote the set of jobs to be processed and p_j is the processing time of job j, $j = 1, \dots, n$. Other parameters include a release date r_j, a deadline d_j , a due date d_j, and a weight w_j. The jobs are partitioned into F families. Let n_j denote the number of jobs in family f, $f = 1, \dots, F$. No setup is required between jobs of the same family. However, the *family setup time* on machine i when a job of family g is immediately preceded by a job of a different family f is s_{ifg}, or s_{i0g} if there is no preceding job. If, for each g, occurs that $s_{ifg} = s_{i0g} = s_{ig}$ for all $f \neq g$, then the setup times on machine i are *sequence independent*; otherwise, they are *sequence dependent*. If, for each machine i, $s_{ifg} = s_{fg}$ for all families f and g including the case $f = 0$, then the setup times are *machine independent*; otherwise, they are *machine dependent*. For the case of a single machine, setup times are, by definition, machine independent. Further, the reasonable assumption is that the *triangle inequality* holds for each machine i, which means that $s_{ifh} \leq s_{ifg} + s_{igh}$, for all distinct families f, g and h, including the case $f = 0$. Unless stated otherwise, the setups are assumed to be *anticipatory*, which means that a setup on a machine does not require the presence of any job. When there are release dates and for shop problems, sometimes the setups allow to be *non-anticipatory*, which means that the setup preceding the processing of some batch cannot start on the current machine before all jobs of this batch are released and have completed their processing on any previous machine.

The minor and mayor setups implementation in a two-stage HFS problem with part family and batch production is proposed in Li (1997). Sequence independent bath setups are considered in Quard & Kuhn (2007), Xuan & Tang (2007). Sequence dependent bath setups are included in the problem described by Luo et al. (2009).

3.5 Burn-in operation

The concept of batch processing is arisen from burn-in operation in semiconductor manufacturing industries which represent today one of the most complex industrial environments. In semiconductor manufacturing, there are parallel machines, different types of processes like batch processes and single wafer processes, sequence-dependent setup times, prescribed customer due dates for the lots, very expensive equipment, reentrant process flows, etc. In such a changeable scenario, maintaining a competitive advantage and remaining profitable in operational terms requires minimization of cycle time, work-in-process (WIP) inventory and the maximization of throughput.

As a part of the complex production line that exists in a semiconductor manufacturing facility, operations involved in BPM are considered to be a bottleneck. This is because the processing times of the lots on the BPM are usually very long compared to other processes, and batching decisions may affect the performance of the entire semiconductor manufacturing process. Semiconductor manufacturing involves numerous batch-processing operations like oxidation, diffusion, deposition, etching, e-beam writing and heat treatment of wafer fabrication, baking of wafer probing, and burn-in operation of device testing.

The *semiconductor burn-in scheduling problem* was first introduced by Lee et al. (2006) and then studied by Mathirajan et al. (2010). The purpose of burn-in operation is to test the integrated circuit (IC) chips. Due to various processes employed in the manufacturing process, some chips may be "fragile" and may fail only after a short period of time. It is essential that these devices are identified and scrapped as "infant mortality". The process of identifying and scrapping these "fragile" devices is known as the *burn-in operation.* It involves subjecting the chips, placed in an oven, to electrical and thermal stress to force the failure of weak or fragile devices.

IC chips arrive at the burn-in area in lots consisting of a number of IC chips of the same product type. Each lot is referred as a job. In a burn-in operation, IC chips of each job are loaded onto *boards*; each job has different lot sizes so that those job sizes (number of demanded boards) are not identical. The boards are often product-specific, and a job cannot be processed without the necessary boards. Once IC chips have been loaded onto the boards, the boards are placed into an oven. Typically, the oven capacity is larger than the job size, so the number of boards in an oven can hold defines the oven capacity, and the size of a job is defined by the number of boards it requires. Each IC chip has a pre-specified minimum burn-in time, which may depend on its type and/or the customer's requirements. Since IC chips may stay in the oven for a period longer than their minimum required burn-in time, it is possible to place different products (jobs) in the oven simultaneously.

The processing time of each batch equals the longest minimum-exposure time among all the products (jobs) in the batch. Effective burn-in operation scheduling is a key issue because it causes frequently a bottleneck due to long processing times relative to other testing operations -e.g. days as opposed to hours- and because it occurs at the end of the manufacturing process and thus has a strong influence on ship dates (Azizoglu & Webster, 2000).

3.6 Cell architectures

The arisen of FMS has triggered many researches dedicated to distinct aspects of their design and functionality. New cell architectures are proposed and new types of management problems are analyzed. Most of the recourse models are dedicated to minimize production costs, that is, the productivity maximization, through optimally allocation and synchronization machinery and parts. Some aspects of those architectures affect the performance of FMS. One of such aspects is the presence of setups.

In several flexible cell architectures each part is mounted on a fixed position in the cell and does not move until the processing machine –i.e. a robot– has completed all the required operations. A FMS recourse (machining center, robot, numerical control center, etc.) needs a certain time to switch the operating mode before the recourse can start (Co et al., 1990; Agnetis et al., 1993; Samaddar et al., 1999; Agnetis et al., 2003). There are generally two different types of setups. One setup occurs when a finished part is removed and replaced with a new part (*part replacement*). Another setup occurs when the machine switches from one operation type to another (*tool switch*). The part replacement setup usually makes more time than the tool switch (Agnetis et al., 2003).

Part replacement may be executed in two different ways named FMS *management policies* (Stecke & Kim, 1988):

- *Batching replacement.* All parts (up to a total k) currently being processed must be completed before new parts are loaded. A setup occurs whenever a new set of parts is loaded (*batch setup*), and another setup occurs at each tool switch (*tool setup*). Parts cannot be removed without stopping the cell operation.
- *Flexible replacement.* Part replacement takes places when it is completed. These setups only occur at tool switches.

Two subproblems must be simultaneously solved (Crama, 1997; Agnetis et al., 2003):

- Forming batches of at most k parts each one.
- Scheduling of the tools (and therefore the robot moves) required by program execution in each batch.

The *tool schedule* is the sequence of tools (possibly repeated) loaded by the machine when processing a given batch. The total time required to process a batch is given by the total duration of the operations, plus the *total tool switching time,* which must be minimized. The length of a tool schedule is the length of the shortest tool schedule for that batch. Hence, for each batch, the scheduling subproblem consists of finding the batch length and the corresponding tool schedule.

Agnetis et al. (2003) provide two illustrative examples of a tool schedule problem for a robot, where any tool schedule is feasible, and provide that it is NP-hard.

Example 1. There are three tools a, b, c, and a batch B consisting of three parts of program P_1 = *abac*, P_2 = *bacb*, and P_3 = *abcb*. For instance, the tool schedule *abcabc* allows to complete P_1 and P_3 but not P_2. Scheduling the tools in the order *abcbacb*, all the part programs can be completed, but this requires six tool switches. On the other hand, the tool schedule *abacb* allows to complete the three part programs using only four tool switches. The length of B is therefore five.

Since the robot performs only one operation at a time, and the total duration of all operations is given, the objective only depends on the amount of time spent in setup activities (both batch and tool setups). The total batch setup time is equal to $T_s\,(b - 1)$, where b is the number of batches. The total setup time is obtained summing the tool setup times of all batches.

Example 2. Besides batch B of the previous example, suppose there is also another batch B' formed by parts P_4 = *bcba*, P_5 = *caba*, and P_6 = *bcab*. The length of B' is also five, obtained for the tool schedule *bcaba*. When batches B and B' are performed, the overall time spent in setup activities is given as $T_s + 8\ t_s$, where T_s is loading new batch setup time, and t_s is any tool switch setup time.

In some situations, the more relevant performance criterion is the number of batches (*switching instants*). This is the case, when the setup time of operations is proportional to the number of tool interchanges, or when the tool transportation system is congested. The distinction between the number of tool switches and the number of switching instants is considered in Tang & Denardo (1988a), Tang & Denardo (1988b) and Sodhi et al. (1994). The more general formulation of the *tool switching problem* is given by Crama (1997) as follows: Determine a part input sequence and an associated sequence of tool loadings such that all the tools required by the j-th part are present in the j-th tool loading and the total number of tool switches is minimized.

Crama et al. (1994) proved that the tool switching problem is NP-hard for any fixed $C \geq 2$, where the number C is the *capacity* of the tool magazine. They also observed that deciding whether there exists a job sequence requiring exactly M tool setups is NP-hard, where M is the total number of tools needed to process all the parts. This latter result conduced to the *gate matrix permutation problem* discussed in the paper of Mohring (1990). The gate matrix permutation problem is to assess the minimum value of the tool magazine capacity such that no tool needs to be set up twice. Since this problem is NP-hard, the tool switching problem is NP-hard too. When all setup times are equal –i.e. when the objective is only to minimize the total number of switches– then the integer program can be solved by a greedy algorithm which turns out to be equivalent to the so-called *keep tool needed soonest policy* (KTNS) –see Tang & Denardo (1988a)–.

4. Lot streaming

4.1 Job splitting

In most multi-stage scheduling studies associated with processing products, a *production batch* (lot) is treated as a single entity called *job* which consists of only one part. As a result, the schedule cannot be improved any more, even if there may be plenty of idle times at machines; that is, partial transfer of completed items in a job between machines is assumed to be impossible. Since the production lots are often large, items already processed on a machine need to wait a long time in the output buffer of this machine whereas the downstream machine may be idle. It can lead to large WIP inventories between the machines and make longer the makespan. When the above assumption is relaxed in the corresponding scheduling problem –i.e. when it is supposed that a job can be split– it may be possible to improve the quality of the resulting schedule. Hence, the production leading times, WIP inventory, interim storage and space requirements, and material handling system capacity requirements can be decreased (Truscott, 1986). The solutions can be implemented in a lesser time and at a lower cost than reorganization. Moreover, when the processing requirement of a job is considered as a total demand of a product in production planning, jobs can be split arbitrarily into continuous sublots and processed independently on m machines to finish the processing all demands as soon as possible (Xing & Zhang, 2000).

As is pointed out by Potts and Van Wassenhove (1992), there are two main advantages of splitting jobs into sublots. Firs, splitting jobs may improve customer service; each sublot can be delivered to the customer immediately upon completion, without waiting for the remaining sublots of the same job. The other motivation for splitting jobs in multi-stage production systems is to enable various operations of the same job to be overlapped by allowing processing of downstream operations to begin immediately for any sublot which has been processed at the current stage.

Baker and Pyke (1990) consider two cases of job splitting:

- Preemption, i.e., Interruption of the production run for a more urgent job;
- Lot streaming, where overlapping operations are permitted.

The concept and practice of lot streaming are not new. The term of lot streaming was first introduced by Reiter (1966). Graves & Kostreva (1986) gave the notion of *overlapping operations* in material requirements planning (MRP) systems. The use of transfer batches (or sublots) is a key element of synchronous manufacturing (Umble & Srikanth, 1990).

So, *lot streaming* is the process of splitting an entire production job (*process batch*) into sublots (*transfer batches*) and scheduling those sublots in an overlapping fashion, in order to accelerate the progress of an order in production. A job is defined here as a production order (*lot*) composed of many identical items (Potts & Baker, 1989), (Baker & Jia, 1993), (Çetinkaya & Duman, 2010). In many practical situations, splitting a lot is both possible and desirable. When a job is split into a number of sublots, a sublot can be processed on a machine even if the other sublots still have not been processed on the upstream machines. The *lot streaming problem* is to decide the optimal number of sublots for each job, the optimal size of each sublot and the optimal sequence for processing the sublots so that the production lead time is minimized. It combines lot sizing and scheduling decisions that were traditionally treated separately (Baker & Jia, 1993; Zhang et al., 2005).

Different sublots of the same job should be processed simultaneously at different stages. As a result of operation overlapping, the production is remarkably accelerated and the idle time on successive machines is reduced. In general, the makespan will be minimized if there is just one item in each sublot (Vickson & Alfredsson, 1992). Nevertheless, there may be practical considerations that make it undesirable to have a large number of unit-sized sublots. It may be possible to attain the minimum makespan with fewer sublots, or there may be difficult in tracking a large number or small sublots. A lot splitting problem finds a compromise between sizes of batch process and sublots when setups are long and difficult. An example of lot streaming benefits for three-machine flow shop, a single job, with the job processing times of 6, 3 and 6 time units is shown on Fig. 1 borrowed from Pan et al. (2010). If the job is not split into sublots, the job completion time is 15 time units (Fig. 1a). When the job is split into three sublots and no-idling production interruption time is allowed between any two adjacent sublots, the job completion time is reduced to 11 time units (Fig. 1b), whereas for the idling case, the job completion time is further reduced to nine time units (Fig. 1c). Obviously, the completion time of the job under the idling case is shorter than the one under the no-idling case with the same sublot type. However, there are also many practical applications for the lot-streaming flow shop scheduling under no-idling case.

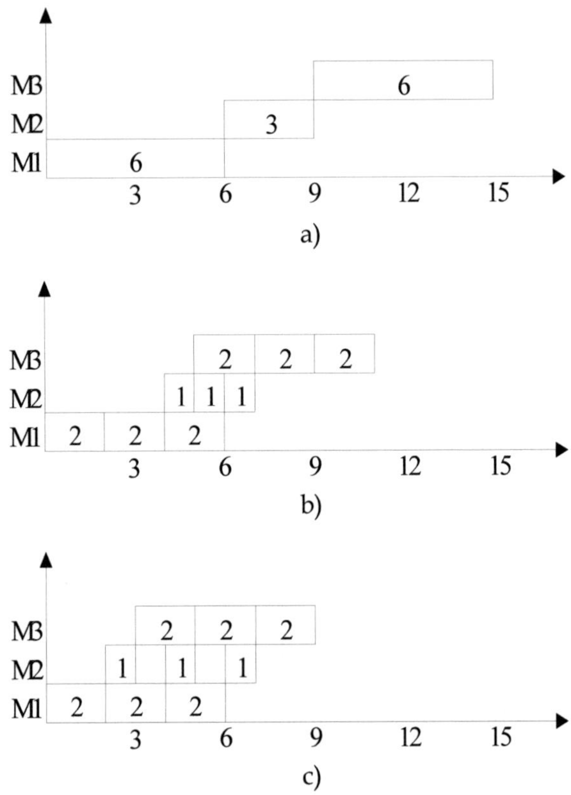

Fig. 1. An example of the lot-streaming flow shop scheduling: a) schedule without sublots; b) schedule with sublots under no-idling case; c) schedule with sublots under idling case.

In the past two decades, with the increasing interest in just-in-time (JIT) and optimized production technology (OPT) philosophies in manufacturing systems, the application of lot streaming idea in scheduling problems has received considerable attention. The JIT approach views each item in the lot as a single unit or job since these results look for a minimum makespan. Through the extensive use of JIT system in manufacturing, the performance measure related to both earliness and tardiness penalties has raised significant attention in lot streaming literature (Kulonda, 1984; Glass et al., 1994; Pan et al., 2010).

One more positive aspect of lot streaming is mentioned in Jeong et al. (1997):

"By assuming that a batch can be split, we can handle the problems that occur due to dynamic nature of shop floor more elegantly. Most past solutions to scheduling problems assume that a manufacturing shop is operating steady and peacefully. But a real factory is full of unexpectedness and dynamics. The causes of such dynamism are very diverse, e.g. human errors, rush orders, requests from customers to shorten the delivery dates, and hardware related events - tool breakage, machine failure, etc. These events cannot be expected beforehand, and therefore, it is very hard to strictly observe the original schedule. As suggested earlier, we can get much better solution considering alternative schedules which allow the splitting of current batches in smaller batches."

Sublot sizes can be (Trietsch & Baker, 1993):

- *Equal* - all sublots of a given lot are of equal size,
- *Consistent* - sublots sizes vary within a lot but are the same for all machines,
- *Variable* - sublot sizes can change from machine to machine.

There are *discrete* and *continuous* versions of lot splitting. In *discrete version* the sublot sizes are integers that correspond to discrete numbers of units in each sublot. Typically, these problems can be formulated as an integer linear program. Anticipating that this may be a difficult problem to solve, lot splitting becomes a *continuous version* of the problem in which the integer restrictions are relaxed. Such a routing may be acceptable if lot size is large and the number of sublots is small. The optimal makespan in the continuous version also serves as a lower bound on the optimal makespan in the discrete version, and the makespan produced by rounding the continuous solution serves as an upper bound (Trietsch & Backer, 1993).

Lot streaming is very common among modern manufacturing systems, furthermore, also introduces an additional complexity in the problems. Each sublot can be viewed as an individual job and the problem size drastically increases with the total number of sublots. If it is assumed that a job consists of a batch as in many real manufacturing environments, then we can obtain an improved schedule. However then, the size of the scheduling problem would become too large to be solved in practical time limit (Jeong et al., 1997). The review on lot streaming problems can be found in several papers –see Potts & Baker (1989), Potts & Wassenhove (1992), Trietsch & Baker (1993) and Chang & Chiu (2005)–.

The publications on lot streaming problems can easily be categorized into two main streams:

- dealing with the determination of the optimal sublot sizes for a single job (also called sublot sizing);
- addressing the sublot sizing and job sequencing decisions simultaneously for the multiple-job case, given various job and shop characteristics.

4.2 HFS lot streaming problem state of art

Lot streaming is very useful as many practical production systems can be considered as HFS, however has received very limited research attention. Some of them will be reviewed in the following.

Tsubone et al. (1996) studied the lot streaming problem in a two-stage HFS with one machine in the first stage and several process lines in the second stage. They examined the impact of lot size, sequencing rules and scheduling scenarios on production makespan, capacity utilization and WIP inventory, using simulation. Zhang et al. (2003) analyzed the integer version of the m-1 HFS lot streaming problem. They solved special cases of the equal-sublot version of the problem, in which one of the stages was obviously a bottleneck. The general problem is formulated as a mixed integer linear programming (MILP) model and solved using two heuristics. Both heuristics enumerated the number of sublots, and for each given number of sublots, allocated the sublots as evenly as possible to stage-1 machines. The sublot sizes were then determined by making them as equal as possible in one heuristic, and by using a smaller MILP model in the other. In this contribution, the continuous version of the problem is studied and efficient optimal solutions to both the problem with given number of sublots and all the cases of the equal-sublot problem is provided. The continuous version does not restrict the sublot sizes to be integers. This is practical in situations where the product is of continuous type –e.g. those in process industries– and where an order consists of a large quantity of small items such as DVDs.

Manufacturing systems considered in studies by Oğuzc et al. (2004), Oğuzc & Frinket (2005) and Ying & Lin (2006) are referred as flow shop with multiprocessors (FSMP). It is a special case of HFS where the parallel machines are assumed to be identical. The overlapping of operations in parallel machines is not considered across stages. Thus, such studies may not be classified as lot streaming research as the very definition of lot streaming is to allow the overlapping of operations across stages.

Zhang et al. (2005) studied the multi-job lot streaming problem in two-stage HFS with *m* identical machines at the first stage and a single machine at the second stage and the objective is to minimize the mean completion time of the jobs. All the *n* jobs are available at time zero. The job sizes are different. A job can be split into sublots that will be treated as separate entities in production. Each sublot requires processing on any one of the machines at the first stage and then on the machine at the second stage. It is assumed that:

There is a given smallest allowable size for the sublots of each job, from which the maximum number of sublots allowed for the job that can be derived.

1. Each machine can process at most one sublot at a time.
2. Each sublot can be processed on at most one machine at a time.
3. The size of each sublot is kept consistent at the two stages.
4. A job can be continuously divisible. As the number of units in ajob is very large, the error between the objective value of the rounded solution and that of the continuous solution may be negligible.
5. The jobs are processed one after another, i.e. the sublots of different jobs are not mixed up in the sequence of processing on any machine.
6. A setup is needed before the processing of each sublot of a job on a machine. The setup times for sublots of the same job at the same stage are equal.
7. The unit processing times and the setup times are known constants.

The problem is then to decide the number and the sizes of sublots for each job and to schedule these sublots to minimize the mean completion time of all jobs. The completion time of a job is defined as the completion time of the last sublot of the job on the second-stage machine.

To solve this NP-hard problem, two heuristics were developed, both using the strategy of first sequencing the jobs and then lot streaming each job. The two heuristics differ in the way of sequencing the jobs. The first heuristic treats each job as a whole lot. The second heuristic considers the system as a pure flow shop with the stage-1 machines aggregated. It uses a profile of each job from the single-job lot streaming result as the time requirements in the artificial pure flow shop. When solving the lot streaming problem of each job in the sequence, both heuristics assign balanced numbers of sublots to the machines at the first stage and decide the sublot sizes using a LP model. A MILP model for the problem was also formulated and used to obtain a lower bound through relaxation. The lower bound was used jointly with two other lower bounds obtained from the direct analysis of the problem structure. Extensive experiments showed that the aggregated machine heuristic performs much better.

Liu (2008) studied the single-job lot streaming problem in a two-stage HFS that has m identical machines at the first stage and one machine at the second stage, called m-1 HFS, the same machine environment that in Zhang et al. (2005). The job is splitted into sublots. It is not necessary for the sizes of the sublots to be equal or to be integers, but they must not be smaller than a lower bound, $x_0 \geq 0$. A sublot will be treated as an independent entity during the production in the system. Each sublot requires processing on anyone of the machines at the first stage and then on the machine at the second stage. The processing times of a sublot at the two stages are proportional to its size. The unit processing times on the stage-1 and stage-2 machines are $p^{(1)}$ and $p^{(2)}$, respectively. Each machine can process at most one sublot at a time. Each sublot can be processed on at most one machine at any time.

Before the processing of each sublot on a machine, a setup time is required for loading the sublot onto the machine. The setup times on the machines of stages 1 and 2 are $s^{(1)}$ and $s^{(2)}$, respectively, which are independent of the processing sequence and the sizes of the sublots. The problem is to determine the number and the sizes of the sublots and the schedule of processing them on the machines to minimize the makespan (the completion time of the entire job).

For the problem with a fixed number of sublots, Liu (2008) proved that it is optimal to use a rotation method for allocating and sequencing the sublots on the machines previously used by Zhang et al. (2003): First, number the sublots is 1, 2, . . ., l. Then, allocate and sequence these sublots on the stage-1 machines in "rotation", i.e. assign sublot 1 to machine 1, sublot 2 to machine 2, ..., sublot m to machine m, sublot $m + 1$ to machine 1, and so on. Finally, sequence the sublots on the stage-2 machine in the order of their numbers.

Hereafter it is referred to the first sublots on the first-stage machines as the first batch of sublots, to the second sublots on these machines as the second batch, and so on. The schedule given by the rotation method is characterized by the following features:

1. The numbers of sublots processed on the stage-1 machines are *balanced* –i.e. each stage-1 machine processes at least $\lfloor l/m \rfloor$ sublots and at most $\lceil l/m \rceil$ sublots–;
2. The sublots are processed on the stage-2 machine in the order of their batch numbers;
3. For the sublots in the same batch, the processing on the stage-2 machine is in the order of their first-stage machine numbers.

When the sublots are allocated using the rotation method, the only remaining decision is to determine the sizes of the sublots. As all the discrete decisions are fixed, the problem of determining the sublot sizes can be formulated as a LP model. Then the problem with equal sublot sizes is considered and an efficient solution to determining the optimal number of sublots is developed. Finally, optimal and heuristic solution methods for the general problem are proposed and the worst-case performance of the equal-sublot solution is analyzed. Computational experiments on a wide range of problem settings, Liu (2008) and Zhang et al. (2003) showed that the heuristic solutions are very close to optimal.

The work of Defersha (2011) was motivated by the gap perceived in research efforts in pure flow shop lot streaming and HFS scheduling. However, the issue of lot streaming in this type of shops has not gained as much attention as for pure flow shop scheduling. Defersha (2011) refers to the work of Zhang et al. (2005) as the only paper that addressed lot streaming in HFS. However, this work is for a very special case where there are parallel machines only in the first stage and the number of stages is limited to two. Lot streaming in a more general HFS with parallel machines on any stage and the number of stages is not limited to two has been studied implicitly and partially.

This chapter is aimed to research in bridging the gap between the efforts in pure flow shop lot streaming and HFS scheduling by presenting a comprehensive mathematical model for lot streaming addressing to the work of Ruiz et al. (2008) which considers the overlapping operations in successive stage through the concept of negative time-lag. This mathematical model incorporates several other practical issues such as unrelated parallel machines, the possibility of certain jobs to skip certain stages, sequence-dependent setup times, anticipatory or non-anticipatory nature of setups, release dates for machines, and machine eligibility. The sublots are to be processed in the order of the stages and sublots of certain products may skip some stages. At a given stage, a sublot of a job can be assigned to one of the parallel machines eligible to process that particular job. For each job there is a sequence dependent setup time on each eligible machine and this setup may be anticipatory or non-anticipatory on different stages. Each machine can process at most one sublot at a time. Sublots of different products can be interleaved. The problem is to determine the size of each sublot of each job, the assignment and processing sequence of these sublots on each machine in each stage. The objective is to minimize the completion time of the last sublot to be processed in the system. The proposed model was solved to optimality for small problem size. The numerical example shown in this referenced paper demonstrated that lot streaming can result in larger makespan reduction in HFS where there is a limited research than in pure flow shop where research is abundant.

5. A problem of demand splitting in a two-stage HFS with lot processing

In this section, an exercise of application of the concepts described trough the section is shown. It consists about the analysis of a real problem settled in electronic components manufacturing; where theoretical formulations as long as modeling results are displayed.

5.1 Production model

The production model is set in reed switch manufacturing. Reed switches are used in electric sensors and relays. The components of a reed switch are two metallic contact blades (reeds) positioned in a hermetically sealed glass tube with a gap between them. The production planning on the plant is realized per lots for a planning horizon, according to the consumer's demands. A demand includes an accepted range of switch operate value named part number, a form of blades, an amount of pieces, a delivery date and needs several lots to be completed.

The technical route of the switch manufacturing is composed of several successive operations divided on a natural manner in two parts with an external operation of one day duration. The investigation is focused on the second part where the presented problem occurs: the classification of pieces and the blade form. The classification operation is realized per lots on one of the group of parallel identical machines, so as one lot is assigned a one of machines. The classification machine measures the value of each piece and deposits it into one of 25 machine repositories according to the value range. Four tube glass types are used in the plant. One lot has a fixed amount of the pieces with the same tube glass type. When a machine starts processing, does not interrupt up to finishing lot. One lot does not divided between machines. There is a list of part numbers. A part number indicates several numbers of repositories so as the same repository can be included in different part numbers. After the classification, a lot is distributed between repositories and then the product is treated in pieces. The content of repositories is assigned on one of four blade form lines taking in account part number and the demand amounts. The rest of pieces not reclaimed for demands forms WIP inventory. The investigated resource model is presented on Figure 2.

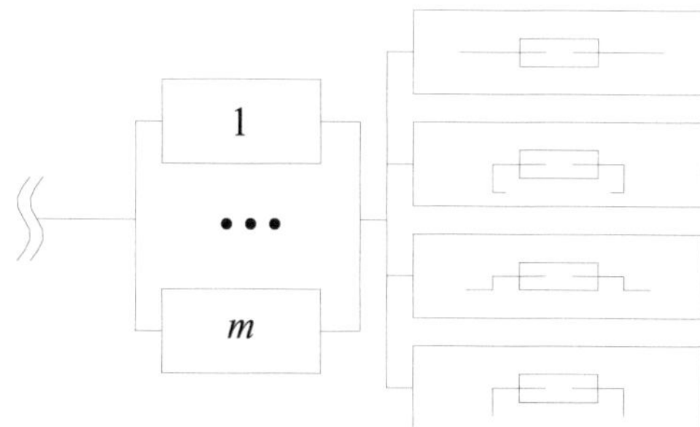

Fig. 2. The *m* identical classification machines and four blade form lines.

The information complexity causes numerous problems in the production planning and scheduling of the plant because they are realized empirically, based on previous experience, and therefore suffer from stochastic results of the lot processing.

5.2 Problem statement

According to the previous experience on the plant, the tube glass type has influence on the distribution of pieces between machine repositories. In (Romero & Burtseva, 2010) is shown that the distribution of pieces belonged to one lot has central tendencies specific for each glass type, and those empirical distributions trend to be near-normal (Fig. 3). A deterministic approach is used here to analyze and resolve the problem, i.e., all distributions are supposed to be known and fixed for each glass type, so as the content of a repository can be anticipatory assigned to a demand.

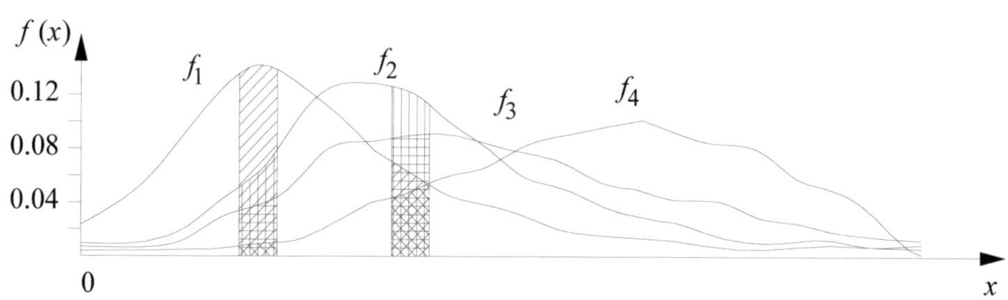

Fig. 3. The probability density functions $f(x)$ of the distributions of the pieces between repositories for different lot glass types.

The follow reasons cause the problem complexity. The quantity of lots for the completion of demands depends essentially on the selection of the glass type of the pieces in the lot in consequence of differences of distributions. Moreover, as the different part numbers often include the same repositories, the allocation of demands on repositories is not evident, e.g., demands 1 and 2 need repositories 1 and 2, and demand 3 needs repositories 2 and 3 (Fig. 4). As a result, the same quantity of pieces for the required part numbers can be obtained from different quantity of lots, and consecutively, the completion time will be different.

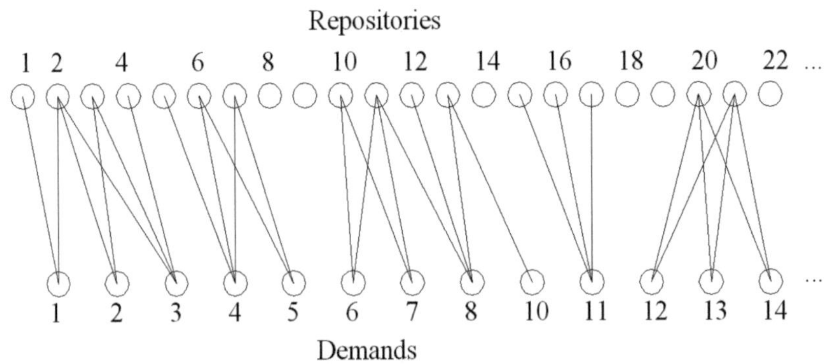

Fig. 4. Bipartited graph of the relationship between repositories and demands (an example).

So, the considered problem is as follows. The n demands must be carried out in two-stage HFS with m_1 parallel identical machines on the first stage and m_2 dedicated machines on the second. The first stage is realized per lots of size U, and the second per batches of sublots. There are G lot types. After first stage, a lot l is distributed into R sublots of size k_{gr} according to the row g of the matrix $K[g,r]$, $g = 1,\ldots,G$, $r = 1,\ldots, R$, i.e., the lot splitting is realized on a natural manner. The demands must be allocated per parts on lot repositories employing the binary chain j of R bits, whose r-element is equal 1 if the repository r is allowed to using for fulfillment of demand j, and is 0 otherwise. In result of the allocation, the sublots s_{lrj} are formed and assigned without mixing to one of m_2 dedicated machines of the second stage. There are sequence independent setup times before the first stage, needed for the allocation of the lot on machine, and between the first and the second stages for the sublots rebatching. The criterion is the makespan minimum.

Using the three-field notation $\alpha\,|\,\beta\,|\,\gamma$, the considered problem can be denoted as:

$$FH2, PM^{(1)}, DM^{(2)}\,|\,\text{lot}^{(1)}, \text{split}^{(1)}, p_l^{(1)} = p, \text{job constraints}^{(1)}, \text{s}_{\text{si}}^{(2)}, \text{batch}^{(2)}\,|\,C_{\max}\,. \qquad (1)$$

Given this situation, the optimization problem addressed in (1) is summarized as follows: The shop model represents a two-stage HFS with a set of parallel identical machines on the first stage and a set of dedicated machines on the second stage. On the first stage, there are lot processing and splitting the lots in form of distribution of pieces into the machine repositories. The lot processing time is constant for any lot. The job constraints on the first stage are mentioned as the restrictions of a job allocation on certain repositories. There is batching of sublots on the second stage. The criterion is the makespan minimum, i.e., the time when the last job is finished.

5.3 Notations

j Demand index, $j = 1,\ldots, N$.
J Demand list, $J = \{j_1, j_2,\ldots, j_N\}$
r Machine repository index, $r = 1,\ldots, R$.
g Lot type, $g = 1,\ldots,G$.
l Lot index, $l = 1,\ldots, L$.
M_1 Set of machines at the first stage, $M_1 = \{1,\ldots, m_1\}$
M_2 Set of machines at the second stage, $M_2 = \{1,\ldots, m_2\}$
V_j Set of repositories associated with job j.
D_j Size of demand j.
U Lot size.
$O = [o_{jr}]$ Utilization of repository r in the job j(part number, N chains of R bits), $O_j = \{o_{j1},\ldots, o_{jR}\}$.
$K = [k_{gr}]$ Matrix of distributions of pieces for a lot of type g into machine repositories r.
L Quantity of lots.
A Resulted lot list, $A = \{g_1, g_2,\ldots, g_l,\ldots, g_L\}$, $g_l \in \{1,\ldots,G\}$.
d_{lrj} Size of sublot s_{lrj}.

5.4 Problem assumptions

- The lots of all types are available at time zero. The size of any lot is U pieces and its processing time is fixed to be $p^{(1)}$. A setup time needed for lot allocating on a machine is included in the lot processing time. The lot selected to processing is identified completely by its glass type g. One lot does not divide between machines. When a machine begins lot processing does not stop until finish.
- The lot l can be assigned to any of m_1 parallel identical machines on the first stage, and next lot is assigned to a machine immediately after finishing of the previous lot. The occupied machines start and finish together. Each machine of the first stage can process at most one lot at time.
- After first stage, the lot of type g is distributed between R machine repositories according to the row g of the matrix $K = [k_{gr}]$; that is, one lot splits into R machine repositories.
- The sublots of size d_{lrj}, $l = 1, 2,\ldots, L$, $r = 1,\ldots, R$, $j = 1,\ldots, N$, are formed using repository contents. A sublot can be formed from only one repository. The content of one repository can be used for one o more sublots.
- The pieces that were not assigned to any job are not used and form the WIP. The WIP content is considered to be used on future demands.
- The sublots belonged to the same job j are joined in batches. The sublots belonged to the different jobs are not mixed. The batches are formed from sublots whose processing is assigned to the same dedicated machine of the second stage.
- There is a non-anticipatory sequence independent setup time s_{in} to form the sublots from the repository contents and the batches of sublots belonged to the same job to be processed on the second stage.
- Each machine of the second stage processes at most one batch at a time. The idle time between batches are permitted on a machine. When a dedicate machine begins processing of a batch does not stop until finish.

5.5 Model of lot batching

To schedule the demand processing, their allocation onto lots must be realized, then, the lots have to be scheduled on the first stage. A demand j can be completed only from certain repositories; formally, it is expressed as follows: The row j of binary matrix $O = [o_{jr}]$ corresponds to the set V_j of repositories associated with demand j. A matrix element $o_{jr} = 1$ if the repository r can be used to complete the demand j. In other cases, $o_{jr} = 0$. Matrixes O and K are employed to find the quantity of pieces of the next demand from the list which can be allocated on the next lot. In result of the allocation, the demand splitting into sublots d_{lrj} and the primary lot batching are realized, then the lot list $A = \{g_1, g_2,\ldots, g_l,\ldots, g_L\}$ of L lots is formed whose elements are the lot types g_l. So, the problem of lots number minimizing is:

$$L = \sum_{g=1}^{G} n_g \rightarrow \min \tag{2}$$

Subject to

$$\sum_{g=1}^{G} \sum_{l_g=1}^{n_g} \sum_{j=1}^{N} d_{l_g rj} \cdot o_{rj} \le \sum_{g=1}^{G} n_g k_{gr}\ , \quad r = 1,\ldots, R, \tag{3}$$

$$\sum_{r=1}^{R} \sum_{j=1}^{N} d_{l_g rj} \le U\ , \quad \forall g, l_g, \tag{4}$$

$$o_{rj} = \begin{cases} 1, \text{ if } r \in V_j \\ 0, \text{otherwise} \end{cases} \quad \forall\, r, j, \tag{5}$$

$$\sum_{g=1}^{G} \sum_{l_g=1}^{n_g} \sum_{r=1}^{R} d_{l_g rj} \cdot o_{rj} = D_j\ , \quad j = 1,\ldots, N, \tag{6}$$

$$k_{gr}, d_{l_g rj} \in Z^0\ , \quad \forall g, l_g, r, j. \tag{7}$$

The constraint (3) describes the relationship between assigned job part sizes and total capacity of repository r, where the total of pieces assigned to a lot l_g must not exceeded its capacity U (4). The binary value o_{rj} in (5) is used to restrict the job part allocation only on the permitted repositories. The equality (6) means that all jobs must be allocated completely. The sizes k_{gr} and $d_{l_g,r,j}$ are the no negative integers (7).

The problem (2) is a generalization of bin packing. In the classical NP-hard bin packing problem one is required to pack a given list of items into the smallest possible number of unit-sized bins. Bin packing has been applied in various areas, e.g.: stock cutting, television programming, transportation, computer storage allocation, bandwidth allocation, scheduling, etc. (Coffman & Csirik, 2007). A difference from classical problem, in the considered case, one lot is associated with a container formed by R sub-containers (bins of machine) (Fig. 5). The bin capacities in a container vary depending on the parameter g (glass type of lot), and are defined according to the pieces distribution given by row g of the matrix K, so that $k_{g1}+k_{g2}+...+k_{gR}$ represents the capacity of the container g (lot size). A demand is associated with an item that corresponds to a set of identical pieces. For each set is indicated the item quantity and bin numbers that are allowed for packing. Detailed survey of the research on the bin packing problem is given by Coffman & Csirik (2007).

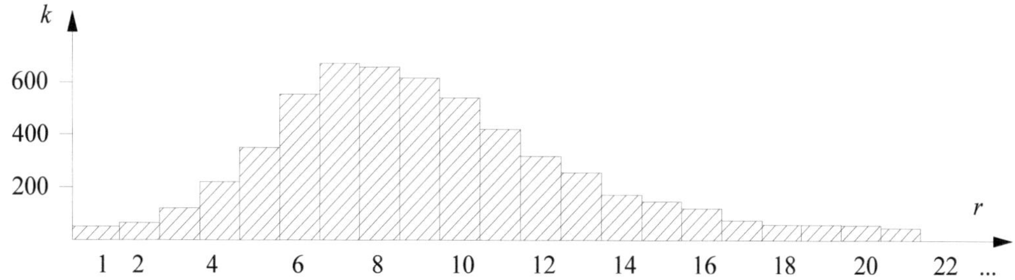

Fig. 5. A container of capacities $k_{g,r}$ formed by R bins,r = 1,…, R,associated with the lot type g.

Shachnai and Tamir (2004) define the problem called the Class-Constrained Bin Packing where the bins have a capacity v and c compartments. In their problem every item has the same size and a color. The items must be deposited on bin, subject to capacity constrains such that items of different colors are placed in different compartments. The goal is to minimize the number of used bins. Fresen and Langston (1986) define the variable sized bin packing problem, where the supply of containers is not only of a single bin type, but some fixed (finite) number of given sizes is available. The bin using cost is simply its size. The goal of the problem is to pack the items into bins witch sum of sizes is minimal. Eptsein and Levin (2008) propose a problem called Generalized Cost Variable Sized Bin Packing. There are given an infinite supply of bins of r types whose sizes are denoted by br < ... < b1 = 1. Items of sizes in (0, 1] are to be partitioned into subsets. A bin type i is associated with a cost c_i, it is assumed c1 = 1. The goal is to find a feasible solution whose total cost is minimized. Langston (1984) investigates the problem of maximizing the number of items packed into m available bins, where the bin sizes can be different. Menakerman and Rom (2001) investigate a bin packing problem variant in which items may be fragmented into smaller size pieces called fragments. Their model is derived from a scheduling problem presented in data over CATV network. Xing (2002) introduces the problem called Bin packing with Oversized Items, where items have a size large than the largest bin size. The bins cannot be overpacked; the oversized item is free to be divided up such that the part is no larger than the largest bin size. Mandal et al. (1998) show that the decision problem for N fragmentable object bin packing when N ≥ 2, is NP-hard. Since this, the problem (2) is NP-hard, too.

5.6 Algorithm

As follows, a heuristic offline algorithm based on the North West Corner rule provides a solution to the problem (2). It finds 1) the quantity of lots indicating for each lot-batch its type and the values of demand parts d_{lrj} allocated on it; 2) the lot sequence.

The algorithm is as follows:

1. Create a table T with N+1 rows and R+1 columns, where each of N rows is used for assignation of pieces of set j, $j = 1,\ldots, N$, to R sub-container. Initially all cells are zero; $n_g = 0$, $g = 1,\ldots, G$. In column R+1 is written the quantity of pieces in each set. The cell on intersection of row j and column r is available if o_{jr}. The unavailable cells are blocked for each row.
2. Sort the rows in the table T using a weight rule (more pieces, fewer pieces, larger number of available sub-container, etc).
3. Create G copies of the table.

   3.1 On the copy T_g, $g = 1,\ldots, G$, add k_{gr} to the value of the cell r on row N+1, $r = 1,\ldots, R$.

   3.2 Process the unblocked cells of the table per rows, starting in the upper left corner while the corresponding value on column R+1 is different from zero. The cell values of row N+1 are assigned to the corresponding cells of row r so that the assigned values do not overflow the value on cell (j, R+1). The assigned value is subtracted from the cells (N+1, r) and (j, R+1).

   It continues until the values of all cells in row N+1 are assigned to the available cells or the value of cell (j, R+1) is zero.
4. Calculate the totals in each table T_g, $g = 1,\ldots G$, as the sum of values on the column R+1.
5. Select the table T_{g^*}, $i \in \{1,\ldots, G\}$, which total is minimum.
   - If totals values in two or more tables are minimal, select the table with minimal total of row $N + 1$.
   - If there is a tie in the two previous rules, select the first table.
     $n_{g^*} = n_{g^*} + 1$.

Keep the packing history (log): the table state T_{g^*} and the selected index g^*.

Add the table T_{g^*} state to the list Q of tables and the index g* to the list W where each element of the list is the number of type g related with the element in sequence of the list Q. The data position for both lists corresponds to the l index.

The quantity $d_{l_g,r,j}$ of pieces of the job j assigned to the repository r is obtained from the table Tg^* located in the l position of the list Q; the g value is located in the l position of the list W.

Clear the available cells for allocation on selected table.

$T_{g^*} \Rightarrow T$.

6. If exist values different from zero on column R+1, go to 2.
7. End

After first stage, the sublots $d_{l_g,r,j}$, $j = 1,\ldots, N$, which processing is assigned on the same machine of the second stage, are coupled without mixing. The lot sequence defines the schedule of these batches on the dedicated machines. For considered real problem is characteristic that processing time of the lot is essentially larger than the second operation duration, independently of the machine number at the first stage –see Fig. 6–, therefore, to optimize the obtained schedule, the lots in the sequence must be arranged in decreasing order of the $\sum_{V_j} \sum_{l_g,r \in V_j} d_{l_g,r,j}$.

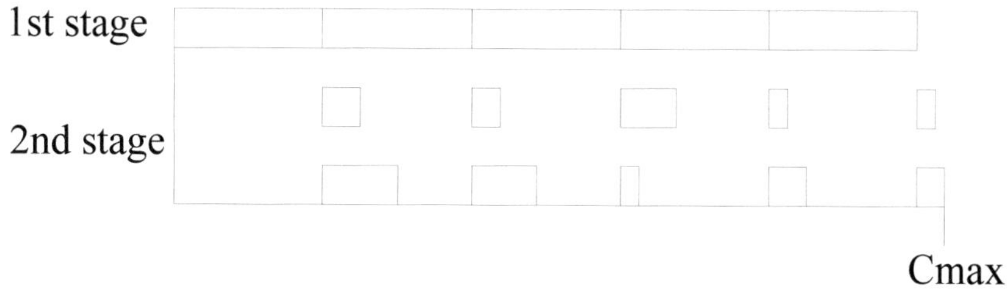

Fig. 6. An example of the problem (2) solution with m_1 =1, m_2 =2.

The next example is presented to illustrate the model working and the algorithm execution: There are 4 demands to carry out (Fig. 7a). For every demand, the next data are indicated: the demanded quantity of pieces, the permitted repositories (in parenthesis), and the assigned dedicated machine (1 or 2). Three lots are necessaries, of types g_1, g_1 and g_2 (Fig. 7b). They arrive on classification machines (Fig. 7c), distributed among machine repositories of the known capacities, and demand are splitted into parts and then are allocated on lots forming the sublots of the sizes $d_{l_g,r,j}$ (batching) (Fig. 7d). These sublots associated with the same demand are joined (Fig. 7e), then sublots which processing is assigned on the same machine of the second stage are coupled (Fig. 7f) (rebatching).

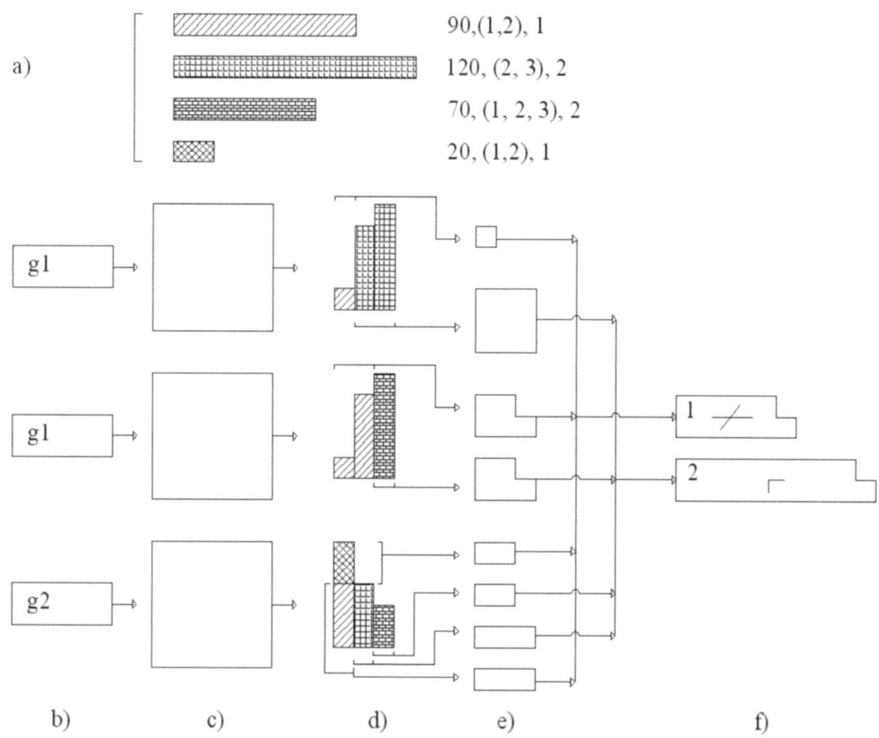

Fig. 7. Model working: a) demands; b) necessary lots indicating lot type; c) first stage machines; d) repository capacities and mode the demands splitting into sublots; e) rebatching sublots.

6. Conclusions

In the last decade, there has been significant interest in scheduling problems that involve lot processing, in consequence of the demand of the modern manufacturing systems. The grouping of jobs into families to process continuously a batch of jobs without any significant setup took to the machines efficiency increasing. But to large batches implicate the large waiting of the operation finish on downstream machines. The splitting of jobs and streaming of lot batches permit to optimize the machine loading to obtain the best schedule. The presented analysis shows that the models in HFS resources environment are insufficiently studied. The concepts described in this work were shown in an example for a real problem of manufacturing electronic components.

7. References

Andrés, C.; Albarracín, J.M.; Tormo, G.; Vicens, E. & García-Sabater, J.P. (2005). Group technology in a hybrid flowshop environment: a case study. *European Journal of Operational Research*, Vol.167, No.1, (November 2005), pp.272-281, ISSN 0377-2217.

Agnetis, A.; Lucertini, M. & Nicolo, F. (1993). Flow management in flexible manufacturing cells with pipeline operations. *Management Science*,Vol.39, No.3, (March 1993), pp. 294-306, ISSN 0025-1909.

Agnetis, A.; Alfieri, A. & Nicosia, G. (2003). Part batching and scheduling in a flexible manufacturing cell to minimize setup costs. *Journal of Scheduling*, Vol.6, No.1, (January-February 2003), pp. 87-108, ISSN 1094-6136.

Azizoglu, M. & Webster, S. (2000). Scheduling a batch processing machine with non-identical job-sizes. *International Journal of Production Research*. Vol.38, No.10, (July 2000), pp. 2173-2184, ISSN 0020-7543.

Baker, K.R. & Pyke, D.F. (1990). Solution procedures for the lot-streaming problem. *Decision Sciences*, Vol.21, No.3, (September 1990), pp. 475–491, ISSN 1540-5915.

Baker, K.R. & Jia, D. (1993). A comparative study of lot streaming procedures. *OMEGA International Journal of Management Sciences*, Vol.21, No.5, (September 1993), pp. 561-566, ISSN 0305-0483.

Bellanger, A. & Oulamara, A. (2009). Scheduling hybrid flowshop with parallel batching machines and compatibilities. *Computers & Operations Research*, Vol. 36, No. 6, (June 2009), pp. 1982-1992, ISSN 0305-0548.

Błażewich, J.; Ecker, K.; Pesch, E.; Schmidt, G. & Węglarz, J. (2007). *Handbook on scheduling: From Theory to Applications*. Springer, ISBN 9783540280460, Berlin, Heidelberg.

Brucker, P.; Gladky, A.; Hoogeveen, H.; Kovalyov, M. Y.; Potts, C. N. & Tautenhahn, T.(1998). Scheduling a batching machine. *Journal of Scheduling*, Vol.1, No.1, (June 1998), pp. 31–54, ISSN 1099-1425.

Brucker, P. (2004). Scheduling Algorithms, Springer, ISBN 3642089070, Osnabru□ck, Germany.

Burbidge, J.L. (1975). The Introduction of Group Technology, *Heinemann Press*, ISBN 0434901938, London.

Chang, J.H. & Chiu, H.N. (2005). A comprehensive review of lot streaming. *International Journal of Production Research*, Vol.43, No.8, (April 2005), pp. 1515–1536, ISSN 0020-7543.

Cheng, T.C.E.; Gupta, J.N.D. & Wang, G. (2000). A review of flowshop scheduling research with setup times. *Production and Operations Management*, Vol.9, No.3, (January 2009), pp. 262–82, ISSN 1937-5956.

Cheng, T.C. E.; Kovalyov, M.Y.; Ng, C.T. & Lam, S.S. (2008). Group sequencing around a common due date. *Discrete Optimization*. Vol.5, No.3, (August 2008), pp. 594-604, ISSN 1572-5286.

Co, H.C.; Biermann, J.S. & Chen, S.K. (1990). A methodical approach to the flexible manufacturing system batching, loading and tool configuration problems. *International Journal of Production Research*, Vol.28, No.12, pp. 2171-2186, ISSN 0020-7543.

Coffman, E. G. Jr. & Csirik, J. (2007). Performance Guarantees for One-Dimensional Bin Packing, In: *Handbook of Approximation Algorithms and Metaheuristics*. Taylor & Francis Group, Chapman and Hall, ISBN 9781584885504, Santa Barbara, USA.

Crama, Y.; Kolen, A.W.J.; Oerlemans, A.G. & Spieksma, F.C.R. (1994). Minimizing the number of tool switches on a flexible machine. *International Journal of Flexible Manufacturing Systems*, Vol.6, No.1, (January 1994), pp. 33-54, ISSN 0920-6299.

Crama, Y. (1997). Combinatorial optimization models for production scheduling in automated manufacturing systems. *European Journal of Operational Research*, Vol.99, No.1, (May 1997), pp. 136–153, ISSN 0377-2217.

Çetinkaya & Duman (2010). Lot streaming in a two-machine mixed shop. *International Journal of Advanced Manufacturing Technology*, Vol.49, No. 9-12, *August* 2010, (*August* 2010), pp. 1161–1173, ISSN 0268-3768.

Damodaran, P. & Srihari, K. (2004). Mixed integer formulation to minimize makespan in a flow shop with batch processing machines. *Mathematical and Computer Modelling*,Vol.40, No.13, (December 2004), pp. 1465–1472, ISSN 0895-7177.

Damodaran, P.; Srihari, K.& Lam, S. (2007). Scheduling a capacitated batch processing machine to minimize makespan. *Robotics and Computer-Integrated Manufacturing*, Vol.23, No.2, (April 2007), pp. 208–216, ISSN 0736-5845.

Defersha, F.M. (2011). A comprehensive mathematical model for hybrid flexible flowshop lot streaming problem. *International Journal of Industrial Engineering Computations*, Vol.2, No.2, (April 2011), pp. 283-294, ISSN 1923-2926.

Dobson, G. & Nambimadon, R. S. (2001). The batch loading and scheduling problem. *Operations Research*, Vol.49, No.1, (January 2001), pp. 52–65, ISSN 1526-5463.

Epstein, L. & Levin, A. (2008). An APTAS for generalized cost variable sized bin packing. *SIAM Journal on Computing*, Vol.38, No.1, (March 2008), pp. 411-428, ISSN 1095-7111.

Friesen,D.K. & Langston, M.A. (1986). Variable size bin packing. *SIAM Journal on Computing*, Vol.15, No.1, (February 1986), pp. 222-230, ISSN 1095-7111.

Glass, C.A.; Gupta, J.N.D. & Potts C.N. (1994). Lot streaming in three-stage production processes. *European Journal of Operational Research*, Vol.75, No.2, (June 1994), pp. 378–394, ISSN 0377-2217.

Graves, S.C. & Kostreva, M.M. (1986). Overlapping operations in material requirements planning. *Journal of Operations Management*, Vol.6, No.3, (May-August 1986), pp. 283–294, ISSN 0272-6963.

Gunasekaran, A.; McNeil, R.; McGaughey, R. & Ajasa, T. (2001). Experiences of a small to medium size enterprise in the design and implementation of manufacturing cells.

International Journal of Computer Integrated Manufacturing, Vol.14, No.2, (March 2001), pp. 212–223, ISSN 1362-3052.

Gupta, J. N. D. & Tunc, E. A. (1998). Minimizing tardy jobs in a two-stage hybrid flowshop. *International Journal of Production Research*, Vol.36, No.9, (September 1998), pp. 2397–417, ISSN 0020-7543.

Hadjinicola, G.C. & Kumar, K. R. (1993). Cellular manufacturing at champion irrigation products. *International Journal of Operations and Production Management*, Vol.13, No.9, pp. 53-61, ISSN 0144-3577.

Jeong, S.; Sangbok, W.; Kang, S.& Park, J. (1997). A batch splitting heuristic for dynamic job shop scheduling problem. *Computers & Industrial Engineering*,Vol.33, No.3-4, (December 1997), pp. 781-784, ISSN 0360-8352.

Jin, Z.H.; Ohno, K.; Ito, T. & Elmaghraby, S.E. (2002). Scheduling hybrid flowshops in printed circuit board assembly lines. *Production and Operations Management*, Vol.11, No.2, (January 2009), pp. 216–230, ISSN 1059-1478.

Kempf, K. G.; Uzsoy, R. & Wang, C. S. (1998). Scheduling a single batch processing machine with secondary resource constraints. *Journal of Manufacturing Systems*, Vol. 17, No. 1, pp. 37–51, ISSN 0278-6125

Kim, J. S.; Kang, S. H. & Lee, S. M. (1997). Transfer batch scheduling for a two-stage flowshop with identical parallel machines at each stage. *OMEGA International Journal of Management Sciences*, Vol.25, No.5, (October 1997), pp. 547–555, ISSN 0305-0483.

Kochhar, S. & Morris, R.J.T. (1987). Heuristic methods for flexible flow line scheduling. *Journal of Manufacturing Systems*, Vol.6, No.4, pp. 299–314, ISSN 0278-6125.

Kulonda, D. J. (1984). Overlapping operations—a step toward just-in-time production. *Readings in Zero Inventory, Proceedings of APICS 27th Annual International Conference*, pp. 78–80, ISBN 0935406514, Las Vegas, NV, USA, October 9-12, 1984.

Kusiak, A. (1987). The generalized group technology concept, *International Journal of Production Research*, Vol.25, No.4, pp. 561–569, ISSN 0020-7543.

Langston, M. A. (1984). Performance of heuristics for a computer resource allocation problem, *SIAM Journal on Algebraic and Discrete Methods*, Vol.5, No.2, (June 1984), pp. 154-161, ISSN 0196-5212.

Lee, C. Y.; Uzsoy, R. & Martin-Vega, L. A. (1992). Efficient algorithms for scheduling semiconductor burn-in operations. *Operations Research*, Vol.40, No.4, (July-Aug 1992), pp. 764–775, ISSN 1526-5463.

Lee, C. Y. & Uzsoy, R. (1999). Minimizing makespan on a single batch processing machine with dynamic job arrivals. *International Journal of Production Research*, Vol.37, No.1, (January 1999), pp. 219–236, ISSN 0020-7543.

Lee, C.-Y.; Leung, J. Y-T. & Yu, G. (2006). Two machine scheduling under disruptions with transportation considerations. *Journal of Scheduling*, Vol.9, No.1, (February 2006), pp. 35-48, ISSN 1099-1425.

Li, S. (1997). A hybrid two-stage flowshop with part family, batch production, major and minor set-ups.*European Journal of Operational Research*, Vol.102, No.1, (October 1997), pp. 142-156, ISSN 0377-2217.

Liao, L. M., & Huang, C. J. (2008). An effective heuristic for two-machine flowshop with batch processing machines. *Proceedings of The 38th conference on computers and industrial engineering (ICCIE2008)*. ISBN 978-7-121-07437-0, Beijing, China, October 31 - November 2, 2008.

Liu, C.Y. & Chang, S.C. (2000). Scheduling flexible flow shops with sequence-dependent setup effects. *IEEE Trans Robotics Automation*, Vol.16, No.4, (August 2000), pp. 408-419, ISSN 1042-296X.

Liu, J. (2008). Single-job lot streaming in m-1 two-stage hybrid flowshops. *European Journal of Operational Research*, Vol.187, No.3, (June 2008), pp. 1171-1183, ISSN 0377-2217.

Luo, H.; Huang, G. Q.; Zhang, Y.; Dai, Q. & Chen, X. (2009).Two-stage hybrid batching flowshop scheduling with blocking and machine availability constraints using genetic algorithm. *Robotics and Computer Integrated Manufacturing*, Vol.25, No.6, (December 2009), pp. 962-971, ISSN 0736-5845.

Lushchakova,I. N.& Strusevich, V. A. (2010). Scheduling incompatible tasks on two machines. European Journal of Operational Research, Vol.200, No.2, (January 2010), pp. 334-346, ISSN 0377-2217.

Manjeshwar, K.; Damodaran, P. & Srihari, K. (2009). Minimizing makespan in a flow shop with two batch-processing machines using simulated annealing. *Robotics and Computer-Integrated Manufacturing*, Vol.25, No.3, (June 2009), pp. 667-679, ISSN 0736-5845.

Mandal, C. A.; Chakrabarti, P. P. & Ghose, S. (1998).Complexity of fragmentable object bin packing and an application. *Computers & Mathematics with Applications*, Vol.35, No.1, (June 1998), pp. 91-97, ISSN 0898-1221.

Marimuthu, S.; Ponnambalam, S.G. & Jawahar, N. (2009). Threshold accepting and ant-colony optimization algorithm for scheduling m-machine flow shop with lot streaming, *Journal of Material Processing Technology*, Vol.209, No.2, (January 2009), pp. 1026-1041, ISSN 0898-1221.

Mathirajan, M. & Sivakumar, A. L. (2006). A literature review, classification and simple meta-analysis on scheduling of batch processors in semiconductor. *International Journal of Advanced Manufacturing Technology*, Vol. 29, No.9-10, (July 2006), pp. 990-1001, ISSN 1433-3015.

Mathirajan, M.; Bhargav, V. & Ramachandran, V. (2010). Minimizing total weighted tardiness on a batch-processing machine with non-agreeable release times and due dates. *The International Journal of Advanced Manufacturing Technology*, Vol.48, No.9-12, (June 2010), pp. 1133-1148, ISSN 1433-3015.

Menakerman, N. & Rom, R. (2001). Bin Packing with Item Fragmentation. *Proceedings of the 7th International Workshop on Algorithms and Data Structures WADS*, LNCS Vol. 2125, ISBN: 9783540424239, pp. 313-324, Providence, RI, USA, August, 8-10, 2001.

Morita, H. & Shio, N. (2005). Hybrid branch and bound method with genetic algorithm for flexible flowshop scheduling problem. *JSME International Journal Series C*, Vol.48, No.1, pp. 46–52, ISSN 1347-538X.

Mohring, R.H. (1990). Graph problems related to gate matrix layout and PLA folding, in: *Computational Graph Theory*. G. Tinhofer et al. (eds.), pp. 17-51, *Springer-Verlag*, ISBN 3211821775, Vienna, Austria.

Mitrofanov, S.P. (1966). *Scientific Principles of Group Technology*. National Lending Library, Yorkshire, UK.

Oğuzc, C.; Zinder, Y.; Do, V. H., Janiak, A. & Lichtenstein, M. (2004). Hybrid flow-shop scheduling problems with multiprocessor task systems. *European Journal of Operational Research*, Vol.152, No.1, (June 2010), pp. 115–131, ISSN 0377-2217.

Oğuzc, C. & Frinket, M. E. (2005). A genetic algorithm for hybrid flow-shop scheduling with multiprocessor tasks. *Journal of Scheduling*, Vol.8, No.4, (July 2005), pp. 323–351, ISSN 1099-1425.

Pan, Q.-K.; Tasgetiren M. F.; Suganthan, P.N. & Chua, T.J. (2010). A discrete artificial bee colony algorithm for the lot-streaming flow shop scheduling problem. *Information sciences*, Vol.18, No.12, (June 2011), pp. 1-14, ISSN 0020-0255.

Perez, I. C., Fowler, J. W., & Carlyle, W. M. (2005). Minimizing total weighted tardiness on a single batch process machine with incompatible job families. *Computers and Operations Research*, Vol. 32, No.2, (February 2005), pp. 327–341, ISSN 0305-0548.

Petrov, V.A. (1966). Flowline Group Production Planning, *Business Publications*, ISBN 0220794715, London.

Pinedo, M. L. (2008). Scheduling: Theory Algorithms, and Systems. *Springer Science+Business Media*, ISBN: 978-0-387-78935-4, NY.

Potts, C.N. & Baker, K.K. (1989). Flow shop scheduling with lot streaming. *Operation Research Letters*, Vol.8, No.6, (December 1989), pp. 297–303, ISSN 0167-6377.

Potts, C.N. & Van Wassenhove, L.N. (1992). Integrating scheduling with batching and lot-sizing: A review of algorithms and complexity. *Journal of the Operational Research Society*, Vol.43, No.5, (May 1992), pp. 395-406, ISSN 0160-5682.

Potts, C. N. & Kovalyov, M. Y. (2000). Scheduling with batching: a review. *European journal of operational research*, Vol.120, No. 2, (January 2000), pp. 228-249, ISSN 0377-2217.

Quadt, D. & Kuhn H. (2007). A taxonomy of flexible flow line scheduling procedures. *European Journal of Operational Research*, Vol.178, No.3, (May 2007), pp. 686–98, ISSN 0377-2217.

Reiter, S. (1966). A system for managing job shop production. *Journal of Business*, Vol.34, pp. 371-393, ISSN 00219398.

Romero Parra, R. & Burtseva, L. (2010). Implementation of Bin Packing Model for Reed Switch Production Planning, In: *IAENG Transactions on Engineering Technologies: Special Edition of the World Congress on Engineering and Computer Science-2009, San Francisco, CA, USA, October 20–22 2009*, Sio-Iong Ao (editor), Vol. 1247, pp. 403-412, AIP Conference Proceedings: ISBN: 978-0-7354-0794-7, Melville, NY.

Ruiz, R.; Serifoglu, F.S. & Urlings, T. (2008). Modeling realistic hybrid flexible flowshop scheduling problems. *Computers & Operations Research*, Vol.35, No.4, (April 2008), pp. 1151-1175, ISSN 0305-0548.

Ruiz, R. & Vazquez-Rodriguez, J. A. (2010). The hybrid flow shop scheduling problem. *European Journal of Operational Research*, Vol. 205, No.1, (August 2010), pp. 1-18, ISSN 0377-2217.

Sammaddar, S., Rabinowitz, G., & Mehrez, A. (1999). Resource sharing and scheduling for cyclic production in a computer integrated manufacturing cell. *Computers & Industrial Engineering*, Vol.36, No.3, (July 1999), pp. 525-547, ISSN 0360-8352.

Santos, D.L., Hunsucker, J.L. & Deal, D.E. (1995). Global lower bounds for flow shops with multiple processors. *European Journal of Operational Research*, Vol.80, No.1, (January 1995), pp. 112-120, ISSN 0305-0548.

Shachnai, H. & Tamir,T. (2004). Tight Bound for online class-constrained packing. *Proceedings of the 6th Latin American Symposium LATIN 2004: Theoretical Informatics*, ISBN 3540212582, pp. 103-123, Buenos Aires, Argentina, April 5-8, 2004.

Sodhi, M.S.; Agnetis, A. & Askin, R.G. (1994). Tool addition strategies for flexible manufacturing systems. *International Journal of Flexible Manufacturing Systems*, Vol.6, No.4, (October 1994), pp. 287-310, ISSN 1572-9370.

Stecke, K.E. & Kim, I. (1988). Study of FMS part type selection approaches for short term production planning. *International Journal of Flexible Manufacturing Systems*, Vol.1, No.1, (September 1988), pp. 7-29, ISSN 1572-9370.

Tang, C.S. & Denardo, E.V. (1988a). Models arising from a flexible manufacturing machine, Part I: Minimization of the number of tool switches. *Operations Research*, Vol.36, No.5, (September - October 1988), pp. 767-777, ISSN 1526-5463.

Tang, C.S. & Denardo, E.V. (1988b). Models arising from a flexible manufacturing machine, Part II: Minimization of the number of switching instants. *Operations Research*, Vol.36, No.5, (September - October 1988), pp. 778-784, ISSN 1526-5463.

Tang, L.; Luh, P.B.; Liu, J. & Fang, L. (2002).Steel-making process scheduling using Lagrangian relaxation. *International Journal of Production Research*, Vol.40, No. 1, (January, 2002), pp. 55–70, ISSN 0020-7543.

Tatikonda, M.V. & Wemmerlov, U. (1992). Adoption and implementation of group technology classification and coding systems: Insights from seven case studies. *International Journal of Production Research*, Vol.30, No.9, pp. 2087–2110, ISSN 0020-7543.

Trietsch, D. & Baker, K.R. (1993). Basic techniques for lot streaming. *Operations Research*, Vol.41, No. 6, (November-December 1993), pp. 1065–1076, ISSN 0030-364X.

Truscott, W. G. (1986). Production scheduling with capacity-constrained transportation activities. *Journal of Operation Management*, Vol.6, No.3-4, (November-December 1993), pp. 333–348, ISSN 0272-6963.

Tsubone, H.; Ohba, H. & Uetake, T. (1996). The impact of lot sizing and sequencing in manufacturing performance in a two-stage hybrid flow shop. *International Journal of Production Research*, Vol.34, No.11, pp. 3037–3053, ISSN 0020-7543.

Umble, M.M. & Srikanth, L. (1990). Synchronous Manufacturing: Principles for World Class Excellence. The Spectrum Publishing Company, Wallingford, CT, USA, ISBN 0-943953-05-7

Uzsoy, R. (1994). Scheduling a single batch processing machine with non-identical job sizes. *International Journal of Production Research*, Vol.32, No.7, pp. 1615–1635, ISSN 0020-7543.

Uzsoy, R. (1995). Scheduling batch processing machines with incompatible job families. *International Journal of Production Research*, Vol.33, No.10, pp. 2685–2708, ISSN 0020-7543.

Vickson, R.G. & Alfredsson, B.E. (1992). Two and three machines flow shop scheduling problems with equal sized transfer batches. *International Journal of Production Research*, Vol.30, No.7, pp. 1551–1574, ISSN 0020-7543.

Wemmerlov, U. & Hyer, N.L. (1989). Cellular manufacturing in the US industry: a survey of current practices. *International Journal of Production Research*, Vol.27, No.9, pp. 1511–1530, ISSN 0020-7543.

Xuan, H. & Tang, L.X. (2007). Scheduling a hybrid flowshop with batch production at the last stage. *Computers & Operations Research*, Vol.34, No.9, (September 2007), pp. 2718–33, ISSN 0305-0548.

Xing W. & Zhang, J. (2000). Parallel machine scheduling with splitting jobs. *Discrete Applied Mathematics*, Vol.103, No. 1-3, (July 2000), pp. 259–69, ISSN 0166-218X

Xing, W. (2002). A bin packing problem with over-sized items. *Operations Research Letters*, Vol.30, No.2, pp. 83-88, ISSN 0167-6377.

Yaurima, V.; Burtseva, L. & Tchernykh, A. (2009). Hybrid Flowshop with Unrelated Machines, Sequence Dependent Setup Time, Availability Constraints and Limited Buffers. *Computers & Industrial Engineering*, Vol.56, No.4, (May 2009), pp. 1452-1463, ISSN 0360-8352.

Yazdani, S.M.T. & Jolai, F. (2010). Optimal methods for batch processing problem with makespan and maximum lateness objectives. *Applied Mathematical Modelling*, Vol.34, No.2, (February 2010), pp. 314-324, ISSN 0307-904X.

Ying, K.-C. & Lin, S.-W. (2006). Multiprocessor task scheduling in multistage hybrid flow-shops: an ant colony system approach. *International Journal of Production Research*, Vol. 44, No.16, (August 2006), pp. 3161–3177, ISSN 0020-7543.

Zhang, W.; Liu, J. & Linn, R. (2003). Model and heuristics for lot streaming of one job in M-1 hybrid flowshops. *International Journal of Operations and Quantitative Management*, Vol.9, (March 2002), pp. 49–64.

Zhang, W.; Yin, C.; Liu,J. & Linn, R. J. (2005). Multi-job lot streaming to minimize the mean completion time in m-1 hybrid flowshops. *International Journal of Production Economics*, Vol.96, No.2, (May 2005), pp. 189-200, ISSN 0925-5273.

Part 3

Heuristic and Metaheuristic Methods

5

Analyzing Different Production Times Applied to the Job Shop Scheduling Problem

Arthur Tórgo Gómez,
Antonio Gabriel Rodrigues and Rodrigo da Rosa Righi
Programa Interdisciplinar de Pós-Graduação em Computação Aplicada,
Universidade do Vale do Rio dos Sinos,
Brazil

1. Introduction

The classic objective of the Job Shop Scheduling Problem (JSSP) is to find a sequence of parts with minimal time to complete all parts (Nowicki and Smutnicki, 1996). The time spent to finalize all parts is known by makespan. In other words, the makespan is the total length of the schedule (when all the jobs have finished processing). Besides the makespan, other objectives can be considered, such as minimize the number of setups, the idle time of machines, the number of tool switches in a machining workstation and so on. The Scheduling Problem is considered hard to solve, with computational complexity defined as NP-Hard (Nowicki and Smutnicki, 1996). It can be applied in a variety of manufacturing systems, being specially studied in Flexible Manufacturing Systems (FMS) (Jha, 1991).

The objective of this study is to show the behavior of three different times in the context of Job Shop Scheduling Problem. The aforementioned times are: (i) tardiness time; (ii) setup time and; (iii) switching tool time. Objective functions were defined with this three production times, representing our decision variables. Cluster Analysis and Tabu Search (TS) Techniques are used to development the model.

The article is divided as follow. Section 2 introduces the concepts of JPSS and its mathematical formulation. Some resolutions methods are mentioned in Section 3. Moreover, Section 4 describes in details the metaheuristic denoted like Tabu Search. This section is responsible for presenting our proposal of a JSSP application. Considering this, this part of the text represents out main contribution regarding the JSSP field. Validation and some experimental results are showed in Section 5. Finally, Section 6 finalizes the chapter, emphasizing its main ideas and contributions.

2. The job shop scheduling problem

One of the classic problems of the area of combinatorial optimization is the Job Shop Scheduling Problem (JSSP), which is defined for minimizing the total production time of a specific system. Generally, JSSP is applied for a range of applications in manufacturing area.

Studied since the 60's, this problem is considered quite complex, and some of its instances waited about two decades to have reached the optimum result.

Since he JSSP comes from the area of Manufacturing, it is common to find variations that reflect different particularities of a production system. In addition, variations also occur when considering different objectives of the minimizing function. Basically, this function is applied to the total production time, but it can also be observed to minimize delivery times, as well as to minimize the number of stops of a machine. JSSP can be treated following two approaches: (i) stochastically and; (ii) deterministic mode. The former works with probability distributions for the arrival of requests for products and processing times, load and displacement within the manufacturing plant. The second approach assumes that the processing times of products are known in advance, as well as both the load times in the machines and the displacement within the factory.

Given the high complexity of JSSP, the accurate methods for solving the optimization combinatorial problems seem to be inefficient and computationally infeasible. This fact is explained because JSSP can demand an enormous amount of time and a high number of computational resources (such as memory and processing power). Therefore, studies involving heuristics and metaheuristics became more and more relevant in this context. Even without guaranteeing an optimal result, they present methods for getting good results with low computational costs. Currently, the JSSP problem serves as benchmark for new metaheuristics being studied by various fields such as engineering and computing. Classically, the Job Shop Scheduling Problem can be defined as a set of parts (or jobs), where each part has associated a set of operations to be performed. Furthermore, there is a set of machines that perform the operations of the aforementioned parts. Once an operation starts, it cannot be interrupted. A classical formulation of this problem is presented below (Blazewicz et al., 1996; Adams et al. 1988; Pezzella and Merelli, 2000):

$$\text{Minimize } tn \tag{1}$$

Subject:

$$tj - ti \geq pi \forall (i,j) \in A \tag{2}$$

$$tj - ti \geq pi \text{ or } ti - tj \geq pj \forall \{i,j\} \in E_k, \forall k \in M \tag{3}$$

$$ti \geq 0 \, \forall i \in V \tag{4}$$

Where, V= {0, 1,...,n} represents the set of operations, where "0" is the first operation and "n" will be the last operation for all jobs. The set of "m" machines is represented by "M" and "A" is the representation of the ordered pairs set of the constraints of operations by the precedence of the relation of each job. For each machine "k" the "Ek" set describes all the operation pairs given by the "k" machine. For each "i" operation, it is processed in a "pi" time (fixed) and the initial "i" process is denoted as "ti" , a variable that has been determinate during the optimization. The Job-shop objective function (1) is used to minimize the makespan. The constraint (2) assures that the sequence of the operation processing for each job corresponds to a pre-determinate order. The constraint (3) assures that there is only one job in each machine at a specific time,. Finally, constraint (4) assures completion of all jobs.

3. Resolutions methods

Many optimization methods have been proposed to the solution of Job Shop Scheduling Problem (Blazewicz et al., 1996) (Mascis and Pacciarelli, 2002)(Zoghby et al., 2004). They can be classified as optimization methods or approximation methods. Considering the optimization methods, we can mention the Integer Programming, Lagrangian Relaxation, Surrogate and Branch and Bound (Balas et al., 1979)(Fisher, 1976). On the other hand, iterative algorithms like Tabu Search, Neural Networks, Genetic Algorithms, Simulated Annealing and GRASP belong to an approach that works with approximation methods (Glover and Laguna, 1997)(Goncalves et al., 2005)(Jain and Meeran, 1998)(Hurink and Knust, 2004). Considering this scenario, Tabu Search is considered a good metaheuristic algorithm for treating problems with a high computational complexity, like JSSP one. Aiming to review some works regarding Tabu Search and others metaheuristics, we recommend some readings (Cordeau et al, 2002; Tarantilis et al, 2005).

3.1 Tabu search

Tabu Search (TS) was proposed by Glover (Glover, 1989) and had its concepts detailed by Glover and Laguna (Glover and Laguna, 1997). Tabu Search is a technique for solving optimization combinatorial problems that consists in iterative routines to construct neighborhoods emphasizing the prohibition of stopping in an optimum local. The main ideas of TS are: (i) It avoids to pass again by recently visited solution area and; (ii) It guides the search towards new and promising areas (Glover, 1986; Wu et al, 2009). Non-improving moves are allowed to escape from the local optima. Moreover, attributes of recently performed moves are stored in a tabu list and may be forbidden for a number of iterations to avoid cycling (Glover, 1986; Wu et al, 2009).

TS searches for the best solution by using an aggressive exploration (Glover and Laguna, 1997). This exploration chooses the best movement for each iteration, not depending on whether this movement improves or not the value of the current solution. In Tabu Search development, intensification and diversification strategies are alternated through the tabu attributes analysis. Diversification strategies drive the search to new regions, aiming to reach the whole search space. The intensification strategies reinforce the search in the neighborhood of a solution historically good (Glover and Laguna, 1997). The stop criterion may be applied to stop the search. It can be defined as the interaction where the best results were found or as the maximum number of iteration without an improvement in the value of the objective function. The tabu list is a structure that keeps some solution's attributes. The objective of this list is to forbid the use of some solutions during some defined time.

4. JSSP aplication

The proposed application considers the classical JSSP with due dates and tooling constraints (Hertz and Widmer, 1996). Each job has a due date in which all its operations shall be completed. Each operation requires a set of tools to processing. The problem is to minimize three production times, represented by decision variables in a objective function *f*. The production times are explained below.

- Makespan (*Ms*): the total time needed to complete the processing of all operations, considering production turns.
- Tardiness time (*At*): the positive difference between the date of completion and the due date of the part, expressed in minutes.
- Setup time (*Sp*): the time spent in preparation for processing new batches during the production of a set of parts, expressed in minutes. This time lasts $\alpha+\beta t$, where α is the time to clean the work area; β is the time to replace one tool and t is the number of tools switched (Hertz and Widmer, 1996; Gómez, 1996).

The managing of the significance of these times is made through the definition of values for the weights of the objective function showed in equation 1.

Considering: d_j is the due date of the job j; x_{ko} is 1 if there is a setup after operation in the machine k, or 0 otherwise;

$$\text{Minimize} \quad W_1 Ms + W_2 At + W_3 St \tag{1}$$

Subject:

$$Ms = \max_{i \in O}(S_i + p_i) \tag{2}$$

$$At = \sum_{j=1}^{J} \max_{i \in O}(S_i + p_i) - d_j \quad \forall \max_{i \in O}(S_i + p_i) - d_j > 0 \tag{3}$$

$$St = \sum_{k=1}^{M} \sum_{o=1}^{M_k} x_{ko}(\alpha + \beta t) \tag{4}$$

$$Ms \geq 0, At \geq 0, St \geq 0, W_1 \geq 0, W_2 \geq 0, W_3 \geq 0. \tag{5}$$

The objective function is expressed by equation (1). Equation (2) represents the Makespan, *i. e.*, the total time to complete the last operation in the schedule. The equation (3) defines the tardiness time, as the sum of differences between the predefined due date (in minutes) and the part completion date. Equation (4) defines the setup time as the sum of all setups of all machines in the schedule. The Equation (5) shows the non-negativity constraints of the decision variables and of the weights.

The proposed application is based on the i-TSAB algorithm developed by Nowicki and Smutnick. (Nowicki and Smutinicki, 2002). It is based on the Tabu Search technique and presents two distinct phases: (i) firstly, the proposed application fills a list E of elite solutions to be explored and; (ii) secondly, using a modified Tabu Search algorithm and the path-relinking technique (Glover and Laguna, 1997), our application explores the solutions and updates E. The modified Tabu Search can reconstruct the best L visited neighborhoods to re-intensification. The algorithm stops when a measure of distance among the solutions in L reaches a threshold. In order to create the neighborhood of feasible results, we are using the Critical Path structure described by Nowicki and Smutniki (Nowicki and Smutnicki, 1996).

The model that represents the application and the proposed objective function was developed in four modules. They are described below.

1. Module 1: Instance Generator - It adapts classical JSSP instances, generating tools for the operations and due dates for the jobs.

2. Module 2: Family of Operations - This module organizes the operations in Family of Operations (FO), according to the tools required for each operation. It implements a Cluster Analysis Algorithm (Kusiak and Chow, 1987; Dorf and Kusiak, 1994).
3. Module 3: Initial Solution - It creates a feasible schedule, ordering the operations O = $1,\ldots,o_j$ $j=1,\ldots n$, $j \in J$.
4. Module 4: Optimization - This module optimizes the initial solution based on the modified i-TSAB technique.

The architecture of the model represented by the information flow among the modules is shown in the Figure 1.

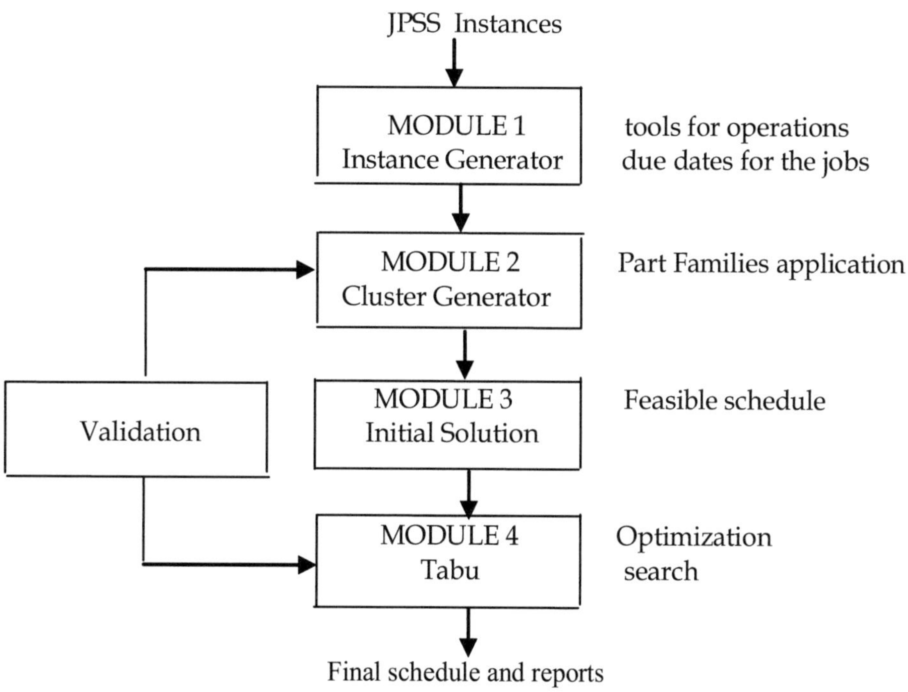

Fig. 1. Architecture of the model that describes the operation of the application

5. Performed experiments

The implementation of the model was made in C language. The source code was compiled by using the GCC compiler, which can be found in the GNU-Linux operational system. The model was validated in two phases: (i) validation of the module 2: generation of Family of Operations (FO) and (ii) validation of module 4: minimization of Ms and St decision variables. Both modules 2 and 4 are illustrated in Figure 1.

5.1 Validation of module 2

The module 2 was validated using a classical instance proposed by Tang and Denardo (Tang and Denardo, 1988). This instance is showed in Figure 2. It is composed by 10 parts and 9 tools. The optimal result for this instance is the generation of 5 Part Families.

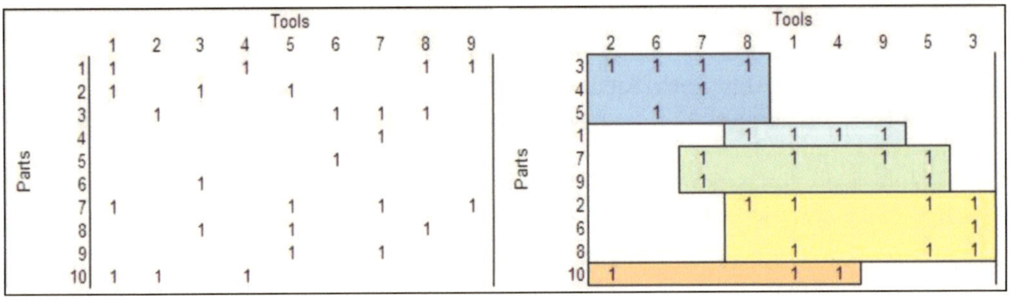

Fig. 2. Instance of 10 parts and 9 tools of Tang and Denardo.

5.2 Validation of module 4

The module 4 generates the schedule through the implementation of the modified i-TSAB technique. Firstly, it was validated the minimization of Setup time comparing results of the module 4 with the work of Hertz and Widmer (Hertz and Widmer, 1996). The authors used 45 benchmark problems provided by Lawrence and Adams *et al* (Jain and Meeran, 1999), adapted to tooling constraints. The search parameters of Tabu Search used in the module 4 were the same used by Hertz and Widmer. In the validation, Module 4 reached the same or better results as Hertz and Widmer. Some results are showed in the Table 1.

Instance	Hertz and Widmer	Module 4
LA16	963	961
LA17	793	789
LA18	876	863
LA19	870	859
LA21	1097	1091
ABZ5	1271	1261
ABZ6	970	963
ABZ7	691	685
ABZ8	701	697

Table 1. Objective function values of Hertz and Widmer and Module 4

The makespan validation was performed using classical JSSP instances proposed by Fisher and Thompson (Jain and Meeran, 1999). These instances are showed in the Table 2. The optimal know result for each instance was reached by Module 4.

Instance	Makespan
FT6	55
FT10	930
FT20	1165

Table 2. Makespan for JSSP benchmark instances.

After validating, there were performed experiments with three minimization politics: (1) minimization of Makespan, (2) Minimization of tardiness and (3) minimization of Setup.

5.3 Benchmark instances and TS parameters

The experiments were performed using 6 benchmark problems provided by Taillard (Taillard, 2006), showed in the Table 3, adapted to the due dates and tooling constraints.

Benchmark instance	Dimensions (job/machine/operation)
$TA1515_1$, $TA1515_2$	15 / 30 / 225
$TA3020_1$, $TA3020_2$	30 / 20 / 600
$TA5015_1$, $TA5015_2$	50 / 15 / 750

Table 3. Instance used in the experiments.

The parameters for the experiments were: production turn lasts 480 minutes; Setup α times lasts 5 minutes; Setup β lasts 4 minutes; machine magazine can hold at most 4 tools; total number of tools needed to process the all operations is 10; any operation requires more than 4 tools for its processing; the Tabu List stores 5 moves; the size of the list of elite solution E is 3; the size of the L list of best visited neighborhood is 1; the measure of distance among solution L is 5.

5.4 Non-tendentious solution

To perform the comparisons among the results obtained with the three minimization politics, it was defined a non-tendentious solution (NTS). It consists in a weight configuration in which any of the decision variables are not privileged. For each instance, it was performed 100 executions of the Module 4, where the values of weights of the objective function were varied in a 0-100 uniform distribution. The mean of the values obtained on each decision variable was calculated and a proportion was made. The Table 4 shows the values obtained for the weights of the decision variables (W_1 (*Ms*), W_2 (*At*) and W_3 (*St*)).

Dimension	W_1	W_2	W_3
15x15	5	1	14
30x20	15	1	29
50x15	21	1	46

Table 4. Weights of the decision variables of f for NTS solution.

Using these values of weights, the modules 3 and 4 were run, for each of the instances. The Tables 5 and 6 show the values obtained for the decision variables with the modules 3 and 4.

Instances	*Ms*	*At*	*St*
$TA1515_1$	10767	73041	2870
$TA1515_2$	10216	78214	2875
$TA3020_1$	27250	393643	7834
$TA3020_2$	27990	390054	7970
$TA5015_1$	33637	777600	10047
$TA5015_2$	31586	734498	9948

Table 5. Values of decision variables of *f* obtained with module 3.

Instances	*Ms*	*At*	*St*
$TA1515_1$	1580	6532	2563
$TA1515_2$	1563	10640	2576
$TA3020_1$	5957	112293	7466
$TA3020_2$	4466	72520	7624
$TA5015_1$	5061	129275	9442
$TA5015_2$	5226	145025	9459

Table 6. Values of decision variables of *f* obtained with module 4.

5.5 Minimization of makespan

In this experiments, the value of the weights of the decision variable *Ms* is increased, meanwhile the weights of the variables *At* and *St* remains the same as the NTS.

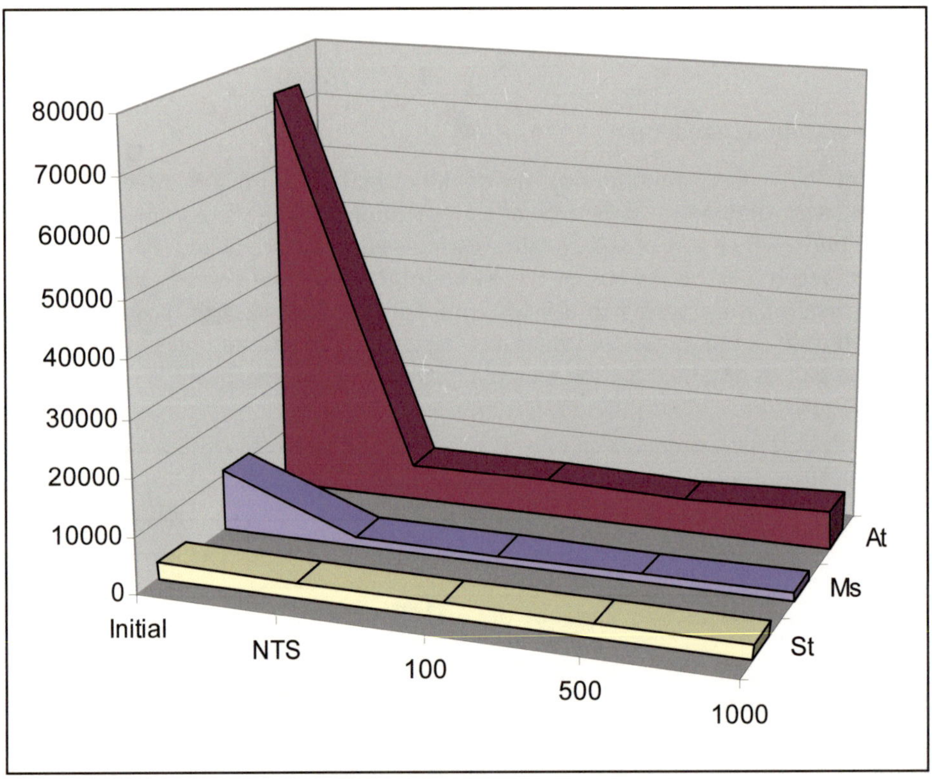

Fig. 3. Values for decision variables of f considering the increasing of Ms weight.

Figure 3 shows the comparison among the initial solution, the NTS and the values obtained with execution of module 4, increasing the value of weights of *Ms*. It can be noticed a reduction of 3% for the *St* decision variable, 87% for the *Ms* and *At* variables, compared to the initial solution. Compared to the NTS solution, the reduction is lower: 3.6% and 6.7% for the *Ms* and *At*. The variable St increases 5.8% compared to the NTS value. This increasing of *St* is due to the fact that the value of this variable depends of the setup in all machines, not only the operations on the Critical Path used to generate neighborhoods.

5.6 Minimization of tardiness

In this experiment, the value of the weight of *At* is increased, while the *Ms* and *St* weights remain the same defined in the NTS previously. The Figure 4 shows the comparison of initial solution, NTS and values obtained with execution of the module 4, increasing the value of the *At* weight.

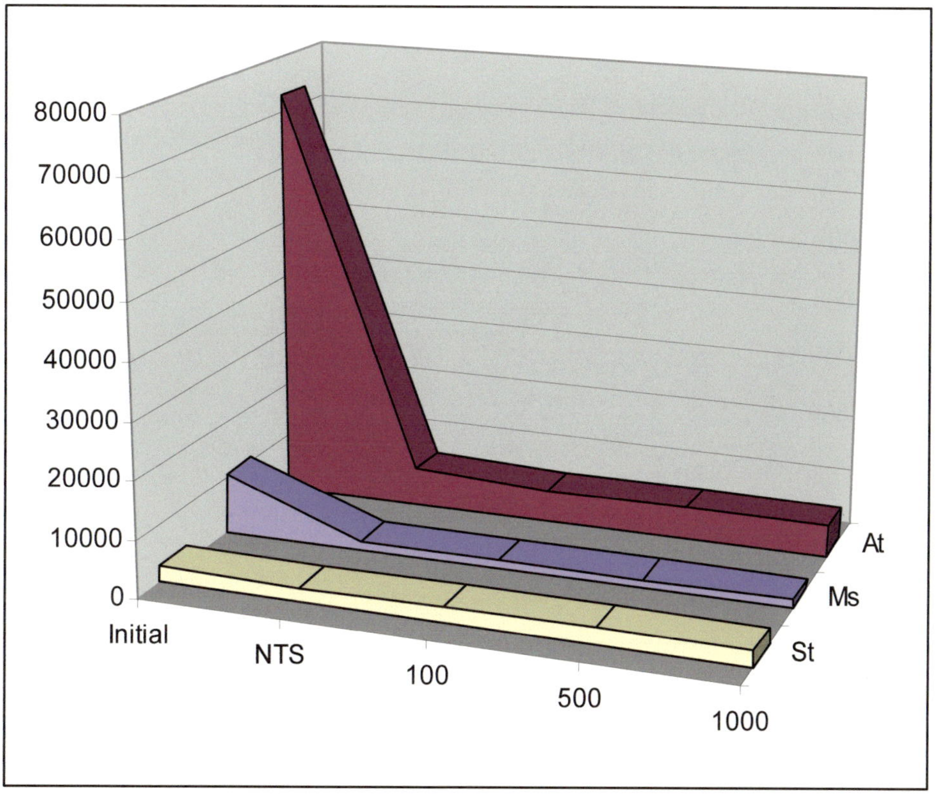

Fig. 4. Values of the decision variables of f considering increasing of At weight.

The increasing of *At* weight reduces 81% the *At* value when compared with the initial solution. The variables *Ms* and *St* had mean reductions of 82% and 2%, respectively. Comparing with NTS values, the values obtained for *At* for instances between 225 and 600 operations are at most 30% lowers. For instances of 750 operations, the *At* assumes higher values (at most 65%). This increasing of the obtained values for *At*, despite the privilege of this variable, is due to the fact that *At* depends of the last operation of each job. Considering that only operations on the Critical Path are moved to generate neighborhoods, *At* variable can be reduced only if the Critical Path contains the last operation of the jobs. Other factor in the comparison between the *At* minimization policy is that NTS reduces significantly the *At* value, compared to the initial solution.

5.7 Minimization of setup

It is considering that the value of *St* weight is increased and the values of *At* and *Ms* weights remains constants, having the NTS values. Figure 5 shows the comparison of the initial solution, NTS and the values obtained with the execution of the Module 4, increasing the value of the *St* weight.

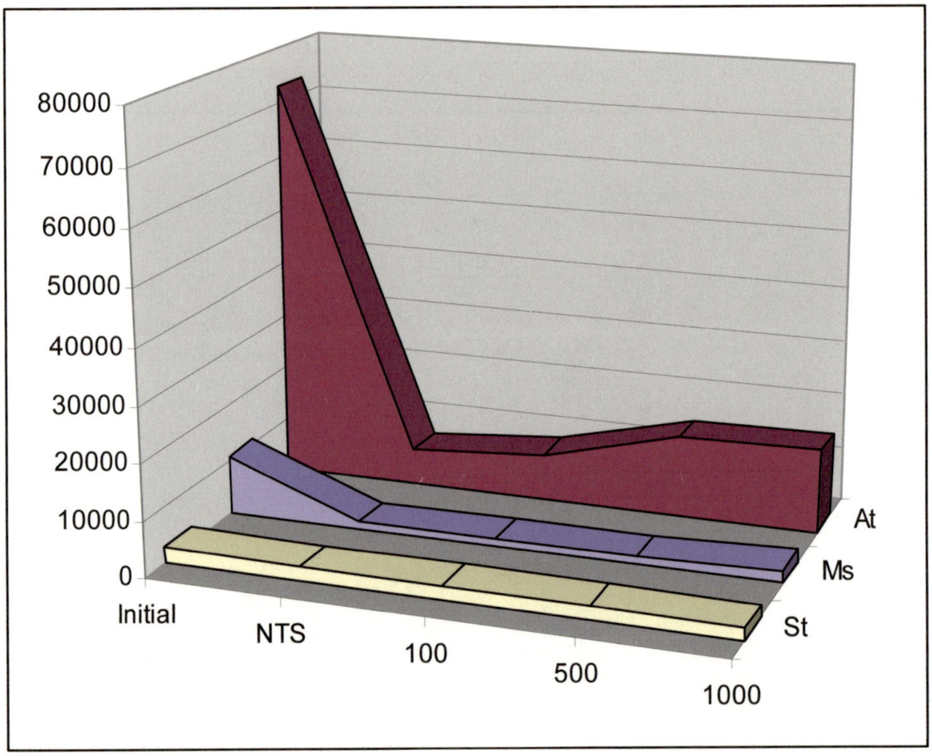

Fig. 5. Values of the decision variable of f considering increasing of St weight.

It can be observed the reduction of the values obtained for *Ms*, *At* and *St* decision variables in 79%, 78% and 14% respectively, compared with the initial solution. When the comparison is made with NTS, the *St* showed reduction of 8.4% in its value. Otherwise, *Ms* increased the obtained value in 24.2% and *At* almost doubled its value.

The setup time always increases when two operations of different FOs are processed in sequence. The machine must stop its processing to switching tools operation. When this operation occurs in the idle time of the machine (e. g. machine is waiting other machine release the product), it not increases the value of *f*. The occurrence of setup is represented, in this work, as a dummy operation that lasts $\alpha+\beta t$ minutes. The total setup time is the sum of all setups in the production, considering that only few of them compose the Critical Path. The setup reduction occurs when two operations of different FOs are swapped in the Critical Path, being operations of the same FO in sequence. Independently of the value assigned to the *St* weight, the operations that not compose the Critical Path will not be considered. The Figure 6 represents the classic instance FT6 adapted to tools constrains. The nodes represent the operations and its processing times. The bold lines represent the operations in the Critical Path. The filled operation represents the setup. It can be noticed that the Critical Path concept does not contribute to the direct minimization of the *St*.

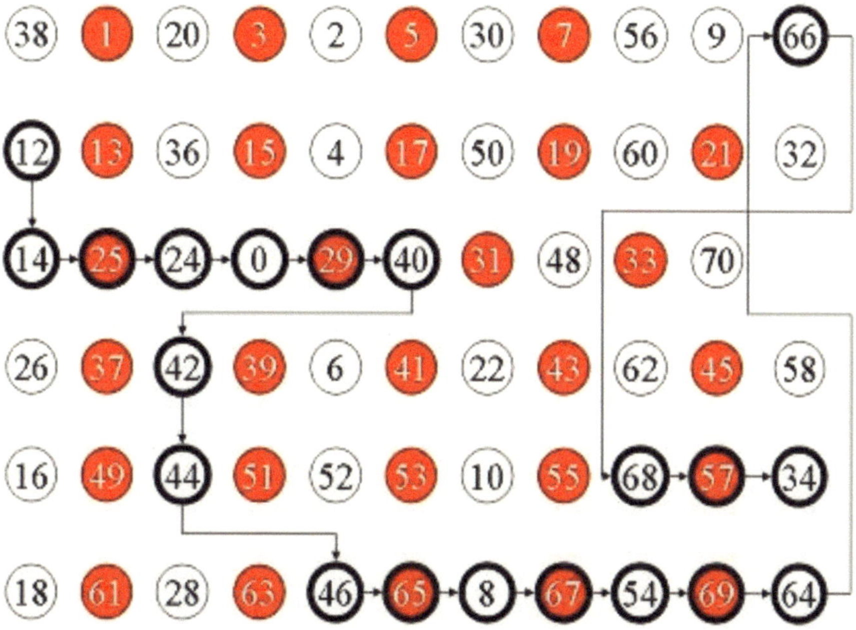

Fig. 6. Graph that represents the FT6 instance.

5.8 TS parameters variation

There were performed experiments where the Tabu Search parameters were changed, with the objective of verify the impact of this change in the generated schedules. Two sets of parameter were used:

1. nbmax = 20,000; Tabu List = 8; L = 5; E = 5.
2. nbmax = 20,000; Tabu List = 15 L = 7; E = 8.

The parameters used in these experiments were: nbmax is 20,000; Tabu List with sizes of 8 and 15; L, list of re-intensification with sizes 5 and 7; E, list of elite solutions with sizes 5 and 8.

The performed experiments were made considering the minimization politics showed in the previous items. There were performed too experiments where each of the decision variables was minimized separately, using the set (2) of parameters. This was performed assigning value 1 for the weight of the decision variable considered and value 0 for the weight of the other two variables.

Figure 7 shows the results of these experiments. The set (3) indicates the minimization of each decision variable separately. The legend *Ms* (1), for example, indicates the minimization of *Ms* policy, using set of parameters (1).

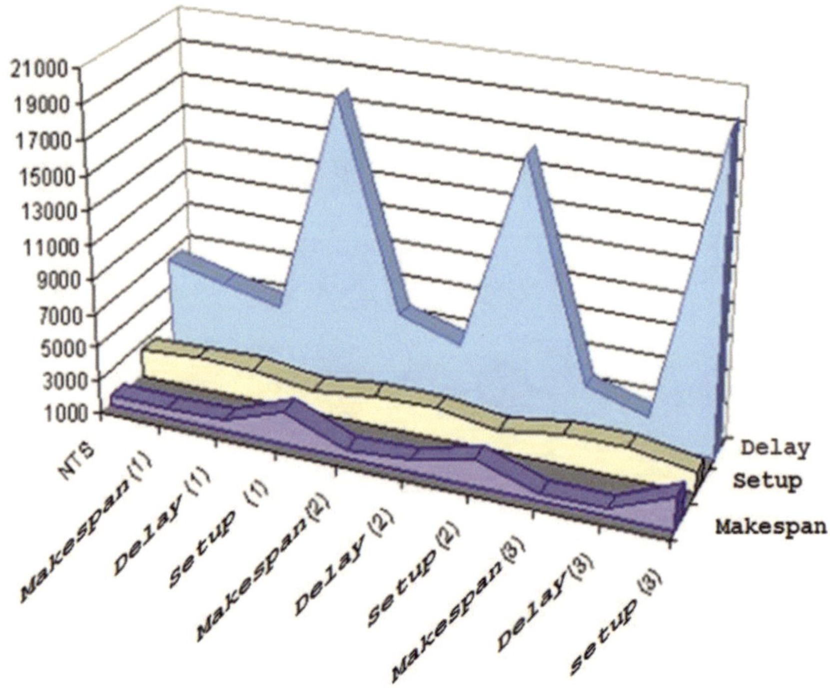

Fig. 7. Comparative graph considering NTS and the minimization politics with the new parameters.

It can be noticed that the increasing of the Tabu Search parameters had positive impact. For each minimization policy, the results obtained with the increased parameters are better than the obtained with NTS. When the decision variables are minimized separately, the reductions obtained for *Ms*, *At* and *St* was 18,4%, 58,1% and 11%, respectively, if compared with NTS. Comparing with initial solution, the reductions were about 88%, 96% and 20%.

In the performed experiments, it can be notices that the implemented model privileges the reduction of *Ms*, due to the use of the Critical Path concept.

6. Conclusion

This chapter proposes a computational model that considers the JSSP problem with due dates, production turns and tooling constraints. The approach used consists in a modification of the Tabu Search technique i-TSAB implemented by Nowicki and Smutinicki (Nowicki and Smutnicki, 2002). The computational model was validated with works of Tang and Denardo (Tang and Denardo, 1988), Hertz and Widmer (Hertz and Widmer, 1996) and using classical JSSP instances provided by Fisher and Thompson (Jain and Meeran, 1999).

There were performed experiments with the objective of verifying the behavior of the model considering three minimization politics: minimization of makespan, tardiness time and setup time. Aiming to compare the results obtained by each policy, it was generated a non-tendentious Solution, in which the three decision variables have the same contribution for the value of *f*.

The implemented model generates good results for the minimization of *Ms*. This is due to the fact that the model is strongly based in the Nowicki and Smutnicki Critical Path concept. The minimization of *At* is not so significant as the *Ms*, considering that the reduction of this variable depends of the lasts operations of the jobs compose the Critical Path. In the *St* minimization, the reduction was not significant too, considering that only setup operations that compose the Critical Path can be reduced in the iterations of the technique.

The increasing of the parameters of the technique had good impact in the search results. A significant reduction of the decision variables can be noticed when each one is minimized separately.

Finally, the proposed model is an approach to the JSSP problem considering additional constraints. The use of the Critical Path concept turns difficult the search for good results when considering the three aforementioned decision variables simultaneously. Thus, the best results were obtained considering the decision variables separately.

7. References

Balas, E., Johnson, E. L., Korte, B., 1979. *Disjunctive programming*. In: Hammer, P.L. Discrete Optimization II, pages 49-55.

Blazewicz, J., Domschke, W., and Pesch, E., 1996. *The job shop scheduling problem: conventional and new solution techniques*. European Journal of Operational Research, 93:1-33.

Dorf, R., Kusiak, A., 1994. *Handbook of Design, Manufacturing and Automation*. John Wiley and Sons.

Fisher, M., 1976. *A dual algorithm for the one-machine scheduling problem*. Math Programming, 11:229-251.

Glover, F.,1989. *Tabu Search – parte1*. ORSA Journal on Computing v.1, n.3.

Glover F., Laguna, M., 1997. *Tabu Search*. Kluwer Academic, Boston.

Gómez, A. T., 1996. *Modelo para o seqüenciamento de partes e ferramentas em um sistema de manufatura flexível com restrições as datas de vencimento e a capacidade do magazine*. PhD Thesis. Instituto Nacional de Pesquisas Espaciais, Brasil.

Groover, M. P., 2001. *Automation, Production Systems, and Computer-Integrated Manufacturing*. United States, Prentice Hall, Second Edition, 856p.

Hertz, A,. Widmer, M.,1996. *An improved tabu search for solving the job shop scheduling problem with tooling constraints*. Discrete Applied Mathematics, 65, 319-345.

Hurink, J., Knust, S., 2004. *Tabu Search algorithms for job-shop problems with a single transport robot*. European Journal of Operational Research, 162:99–111.

Jain, A. S., Meeran, S., 1998. *Deterministic job-shop scheduling: Past, present and future*. European Journal of Operational Research, 113(2):390–434.

Jha, N. K., 1991. *Handbook of Flexible Manufacturing Systems*. Academic Press, 328p.

Kusiak, A., Chow, W. S., 1987. *Efficient Solving of the Group Technology Problem*. Journal of Manufacturing Systems, 6(2):117-124.

Mascis, A., Pacciarelli, D., 2002. *Job-shop scheduling with blocking and no-wait constraints*. European Journal of Operational Research, 143:498–517.

Nowicki, E., Smutnicki, C., 1996. *A fast tabu search algorithm for the permutation flow shop problem*. European Journal of Operational Research. Elsevier, v.91, p.160–175.

Nowicki, E., Smutnicki, C., 2002. *Some new tools to solve job shop problem.* Institute of Engineering Cybernetics, Wroclaw University, Poland.

Taillard, E. 2006. *Job Shop Scheduling Problem on-line instances. Available* in: http://ina2.eivd.ch/collaborateurs/etd/problemes.dir/ordonnancement.dir/ordonnancement.html.

Tang, C. S., Denardo, E. V., 1988. Models arising Flexible Manufacturing Machine Part II: minimization of the number of switching instants. Operations Research, 36(5).

Wu, T., Yeh, J., Chang, C., 2009. *A hybrid Tabu Search Algorithm to Cell Formation Problem and its Variants*. World Academy of Science, Engeneering and Technology, v. 53, 1090-1094.

Zoghby, J., Barnes, W. L., Hasenbein, J. J., 2004. *Modeling the Reentrant job shop problem with setups for metaheuristic searches.* European Journal of Operational Research, 167:336–348.

6

Adaptive Production Scheduling and Control in One-Of-A-Kind Production

Wei Li and Yiliu Tu
The University of Calgary, Canada

1. Introduction

Mass customization is one of competitive strategies in modern manufacturing (Blecker & Friedrich, 2006), the objective of which is to maximize customer satisfaction by producing highly customized products with high production efficiency. There are two starting points moving towards mass customization, mass production and one-of-a-kind production (OKP). The production volume of mass production is normally large, whereas that of OKP is usually small or extremely even just one. Mass production can achieve high production efficiency but relatively low customization, because products are designed in terms of standard product families, and produced repetitively in large volume. Comparatively, OKP can achieve high customization but relatively low production efficiency, because product design in OKP is highly customer involved, and each customer has different requirements. Therefore, the variation of customer requirements causes differences on each product. To improve production efficiency, OKP companies use mixed-product production on a flow line (Dean et al., 2008, 2009). Moreover, the production scheduling and control on OKP shop floors is severely challenged by the variation of customer requirements, whereas that in mass production is comparatively simple. Therefore, we focus on the adaptive production scheduling and control for OKP.

1.1 Characteristics of one-of-a-kind production

OKP is product-oriented, not capacity-oriented (Tu, 1996a). Customers can only choose a product within one of product families provided by an OKP company. Although customer choice is confined by product families, OKP is so customer involved that every product is highly customized based on specific customer requirements, and products differ on matters of colors, shapes, dimensions, functionalities, materials, processing times, and so on. Consequently, production of a product is rarely repeated in OKP (Wortmann et al., 1997). Moreover, OKP companies usually adopt a market strategy of make-to-order or engineering-to-order. Therefore, it is very important to meet the promised due dates in OKP. This market strategy challenges production scheduling and control differently from that of make-to-stock.

Typically, there are five types of problems challenging production scheduling and control in an OKP company. (1) Job insertion or cancellation frequently happens in OKP due to high customer involvement. (2) Operator absence or machine breakdown needs to be carefully controlled to fulfill the critical due dates. (3) Variation in processing times usually happens to an operation, because a highly customized product is rarely repeated. (4) The overflow of work-in-process (WIP) inventories occurs. (5) Production delay on the previous day will affect the production on the current day; so will production earliness. When these problems dynamically happen to an OKP company, the daily production has to be adjusted online, i.e. adaptive production control. Therefore, OKP companies are continuously seeking new methods for adaptive production scheduling and control on shop floors.

1.2 Former research of flow shop production scheduling and control

Flow shop production scheduling has been researched for more than five decades since 1954 (Gupta & Stafford, 2006). Early research of flow shop production scheduling was highly theoretical, using optimization techniques to seek optimal solutions for *n*-job *m*-machine flow shop scheduling problems. However, the emergence of NP-completeness theory in 1976 (Garey et al., 1976) profoundly influenced the direction of research in flow shop production scheduling. NP-completeness implies that it is highly unlikely to get an optimal solution in a polynomially bounded duration of time, for a given complex problem in general. That is why heuristics are required to solve large problems.

Adaptive production control acutely challenges the research of flow shop production scheduling, because the relationship has not been completely revealed, among the number of jobs, the number of machines, job processing times and scheduling objectives. Moreover, the research of flow shop production scheduling is often based on strong assumptions, such as no machine breakdown or operator absence, processing times and some constraints are deterministic and known in advance (MacCarthy & Liu, 1993). During real production, disturbances are manifested in such occurrences as machine breakdown, operator absence, longer than expected processing times, new emergent orders, and so on (McKay et al., 2002), all of which may fail the original offline schedule and then require online re-scheduling for adaptive production control. Consequently, heuristics based on strong assumptions are not robust, making production scheduling systems inflexible (Kouvelis et al., 2005), and a large gap exists between theoretical research and industrial applications (Gupta & Stafford, 2006; MacCarthy & Liu, 1993).

1.3 Status of production scheduling and control in OKP

Currently, OKP companies primarily use priority dispatching rules (PDRs) to deal with disturbances. It is fast and simple to use PDRs to control production online, but PDRs depend heavily on the configuration of shop floors, characteristics of jobs, and scheduling objectives (Goyal et al., 1995), and no single specific PDR clearly dominates the others (Park et al., 1997). Moreover, the performance of PDRs is poor on some scheduling objectives (Ruiz & Maroto, 2005), and inconsistent when a processing constraint changes (King & Spachis, 1980). Consequently, there is a considerable difference between the scheduled and actual production progress (Ovacik & Uzsoy, 1997), and production may run into an "ad hoc fire fighting" manner (Tu, 1996a, 1996b).

Here is a real situation in Gienow Windows and Doors, Canada. Without a computer-aided system for adaptive production scheduling and control, an experienced human scheduler in Gienow carries out scheduling three days before the real production. It is an offline scheduling. Processing times of operations are quoted by Gienow's standards, which are the average processing times of similar operations in the past. On the production day, the production is initially carried out according to the offline schedule. However, real processing times of highly customized products might not be exactly the same as the quoted ones. Therefore, customer orders may be finished earlier or later than they are scheduled offline. This will cause problems such as the overflow of WIP inventories, the delay of customer orders, and so on. The production delay of customer orders is not allowed in Gienow, because the delivery schedule has a high priority. In addition, unexpected supply delays, machine breakdown and operator absence could even cause more problems. To cope with these dynamic disturbances, the shop floor managers and production scheduler in Gienow carry out the following activities based on their experience:

1. Re-allocate operators among work stages in a production line or lines.
2. Change the job sequence.
3. Postpone the production of other orders purely for a rush order
4. Cancel or insert orders into the current production.
5. Alter the production routine to divert orders from one production line to another.
6. Add more work shifts or overtime working.

Carrying out these activities by experience may avoid the overflow of WIP inventories in one stage or line, but cause it in other stages or lines, smoothing the production progress in one stage but slowing down the whole progress in Gienow. Due to the lack of an efficient computer system, Gienow does the adaptive production scheduling and control manually and inefficiently. Obviously, OKP shop floors have to be adaptively scheduled and controlled by a computer aided system (Wortmann et al., 1997; Tu, 1996b).

The rest of this chapter is organized as follows. Section 2 gives a brief literature review on flow shop production scheduling. Section 3 introduces a computer-aided production scheduling system for adaptive production scheduling and control in OKP, consisted of a feedback control scheme and a state space (SS) heuristic. Section 4 gives the results of various case studies. Finally, section 5 draws conclusions and proposes future work.

2. Literature review

In this section, we briefly review research of flow shop production scheduling from two perspectives first, seeking optimal solutions and seeking near-optimal solutions, and then discuss the requirements of heuristics for adaptive production scheduling and control.

2.1 Flow shop scheduling

2.1.1 Definition of flow shop scheduling

Scheduling is a decision making process of allocating resources to jobs over time to optimize one or more objectives. According to Pinedo (2002), one type of flow shop consists of m machines in series, and each job has the same flow pattern on m machines. This is typically called a traditional flow shop (TFS). Another type of flow shop is called a flexible flow shop or hybrid flow shop (HFS), where there are a number of machines/operators in parallel in each of S stages. In addition to the difference of flow shop configurations, processing constraints are also different for TFS and HFS. For TFS, if the first in first out (*FIFO*) rule is applied to jobs in WIP inventories, it becomes a no pre-emption flow shop problem. It is also called a permutation (*prmu*) flow shop problem, because the processing sequence of jobs on each machine is the same. For HFS, because there are multiple machines/operators in a stage, the first job coming into a stage might not be the first job coming out of the stage. Therefore, the first come first serve (*FCFS*) rule is applied (Pinedo, 2002). Consequently, it is still a problem of no pre-emption flow shop. Another processing constraint could be no waiting (*nwt*), that is, there is no intermediate storage or WIP inventories between two machines or stages. The most common objective of flow shop scheduling is to minimize the maximum completion time or makespan, i.e. min(C_{max}). By the three parameter notation, $a/\beta/\gamma$ (Graham et al., 1979), the above problems can be notated as $Fm/prmu/C_{max}$ for m machine TFS problems with no pre-emption to minimize makespan, $Fm/nwt/C_{max}$ for m machine TFS problems with no waiting, $FFs/FCFS/C_{max}$, for S-stage HFS problems with $FCFS$, and $FFs/nwt/C_{max}$ for S-stage HFS problems with no waiting.

2.1.2 Research of flow shop scheduling for optimal solutions

2.1.2.1 Johnson's algorithm

Johnson proposed his seminal algorithm to get optimal solutions for n-job 2-machine flow shop problems in 1954 (Johnson, 1954), the objective of which is to min(C_{max}). The mathematical proof of his algorithm by using combinatorial analysis is as follows.

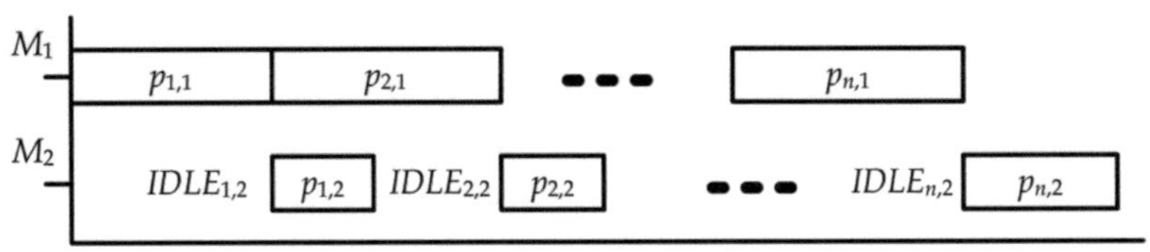

Fig. 2.1 n-job 2-machine flow shop problems, to min(C_{max})

The makespan or C_{max} consists of the sum of processing times and the sum of idle times caused by n jobs on the last machine (Fig. 2.1). For n-job 2-machine flow shop problems, $C_{max} = \sum_{i=1}^{n} p_{i,2} + \sum_{i=1}^{n} IDLE_{i,2}$. The sum of processing times of n jobs on the last machine is a constant. Thus, the objective to min(C_{max}) is converted to minimize the sum of idle times on the last machine. Johnson models the sum of idle times caused on machine 2 as $\sum_{i=1}^{n} IDLE_{i,2} = \max_{1 \le u \le n} \{K_u\}$, where $K_u = \sum_{i=1}^{u} p_{i,1} - \sum_{i=1}^{u-1} p_{i,2}$, in which $p_{i,1}$ and $p_{i,2}$ are the processing times of job i on machine 1 and machine 2 respectively.

To illustrate how to sequence n jobs, Johnson uses a combinatorial analysis approach, which is to compare two sequences, $\{\rho, i, i+1, \pi\}$ and $\{\rho, i+1, i, \pi\}$. The main difference of the two sequences is that two jobs exchange the positions, and ρ is a subset for selected jobs, π for unselected jobs, $\rho \cap i \cap i+1 \cap \pi = \emptyset$, and $\rho \cup i \cup i+1 \cup \pi = \{n\}$. An optimal ordering of jobs is given by the following scheme. Job i proceeds job i+1, if max $\{K_{1u}, K_{1u+1}\} \leq$ max $\{K_{2u}, K_{2u+1}\}$. By subtracting $\sum_{i=1}^{u+1} p_{i,1} - \sum_{i=1}^{u-1} p_{i,2}$ from every term of equation in the above scheme, we can get min $\{p_{i,1}, p_{i+1,2}\} \leq$ min $\{p_{i+1,1}, p_{i,2}\}$, and Johnson's algorithm (JA) is developed accordingly.

2.1.2.2 Extension of combinatorial approach

Dudek and Teuton extend Johnson's combinatorial approach to n-job m-machine flow shop problems to min(C_{max}) (Dudek & Teuton, 1964), comparing the same two sequences as in Johnson's proof, and then develop their dominance conditions. Dudek and Teuton began the analytical framework for the development of dominance conditions for flow shop scheduling, although their initial method is shown to be incorrect later (Karush, 1965).

Smith and Dudek correct Dudek and Teuton's combinatorial approach, by introducing partial enumeration into dominance conditions (Smith & Dudek 1967). They propos two checks of dominance conditions. One is job dominance check and the other is sequence dominance check. The job dominance checks two different sequences, $\{\rho, i, i+1, \pi', \pi''\}$ and $\{\rho, i+1, \pi', i, \pi''\}$, in which π' and π'' are all possible combinations of exclusive subsets of π. The sequence dominance checks another two sequences, $\{\rho, \pi\}$ and $\{\rho', \pi\}$, in which ρ and ρ' are different permutations of the same selected jobs. The two dominance checks theoretically guarantee the optimal solution, but practically are still time consuming.

Based on D-T's framework, Szwarc proposes an elimination rule different from S-D's dominance checks (Szwarc, 1971a, 1971b). Let t $(\rho a, k)$ be the completion time of all jobs of sequence ρa on machine M_k. Then $t(\rho a, k) = \max\{t(\rho a, k\text{-}1), t(\rho, k)\} + p_{a,k}$ with $t(\emptyset, k) = t(\rho, 0) = 0$, where $k = 1,\dots,m$. Define the difference of completion times of two sequences as $\Delta_k = t(\rho ab, k) - t(\rho b, k)$, for $k = 2,\dots,m$. The elimination rule is to eliminate all sequences of the form ρb if $\Delta_{k-1} \leq \Delta_k \leq p_{a,k}$. However, Szwarc clearly stated that "*if there is no job c such that for all k: $c_1 \leq c_k$ or $c_m \leq c_k$, then no single sequence could be eliminated. In this case, the elimination method offers no advantage since we could have to consider all n! sequences*".

2.1.2.3 Branch and bound methods

Besides the combinatorial approach, a branch and bound (BB) method is also a general framework for NP-hard problems. It can be used to get optimal solutions to flow shop scheduling problems (Ignall & Schrage, 1965; Lageweg et al., 1978).

Usually, there are mainly three components in a BB method, a search tree, a search strategy, and a lower bound. A search tree represents the solution space of a problem (Fig. 2.2), the nodes on the tree represent subsets of solutions, and the descendants or child-nodes are given by a branching scheme. For an n-job m-machine flow shop problem, the search tree begins with a virtual node 0. For the first position in a sequence, there are n candidates or nodes, i.e. each of n jobs can be a candidate for position 1. If one job is selected for position 1, it will have n-1 descendants or child-nodes. Consequently, there are $n\times(n-1)$ nodes for position 2, $n\times(n-1)\times(n-2)$ nodes for position 3, and finally, $n!$ nodes for the last position n.

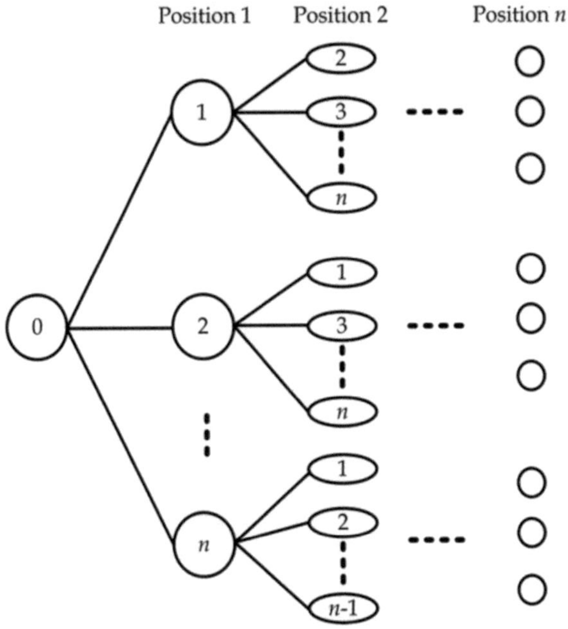

Fig. 2.2 A solution space of a BB method

At each node, a lower bound is calculated in terms of makespan for all permutations that descend it. For each position, all nodes are examined and a node with the least lower bound is chosen for branching. When a node represents an allocation of all jobs and has a makespan less than or equal to the lower bound, it is an optimal solution.

2.1.3 Heuristics for near-optimal solutions

Framinan et al. propose a general framework for the development of heuristics (Framinan et al., 2004). It has three phases: index development, solution construction and solution improvement. Phase 1, index development, means a heuristic arranges jobs according to a certain property of processing times. For example, Campbell et al. propose a CDS heuristic for an n-job m-machine TFS problem to min(C_{max}) (Campbell et al., 19770). CDS arranges jobs as follows. If there is a counter (*Ctr*) pointing to a machine j, then for each job i ($i = 1,\ldots,n$) the sum of processing times on the first *Ctr* machines is regarded as its processing time on virtual machine 1, and that on the rest m–*Ctr* machines as on virtual machine 2. Then apply JA to this virtual 2-machine flow shop problem to get a sequence. As *Ctr* changes from machine 1 to machine m–1, m–1 sequences are generated by CDS, and the one with the minimum makespan is the final solution. In phase 2, solution construction, a heuristic constructs a job sequence by a recursive procedure, trying to insert an unscheduled job into a partial sequence until all jobs are inserted. NEH heuristic (Nawaz et al., 1983) is a typical heuristic in phase 2, for an n-job m-machine TFS problem to min(C_{max}). NEH constructs a job sequence as follows. Step 1, NEH heuristic calculates the sums of processing times on all of m machines for each of n jobs, and then arranges these sums in a non-ascending order. Step 2, NEH heuristic schedules the first two jobs in the order to get a partial sequence. Step 3, NEH heuristic inserts the third job into three possible positions to get another partial sequence, and so on. Finally, NEH heuristic inserts the last job into n possible positions, and then determines the final sequence. In phase 3, solution improvement, heuristics have two main characteristics, an initial sequence generated by other heuristics and artificial intelligence to improve the initial sequence. One typical heuristic in phase 3 is an iterated greedy (IG) heuristic (Ruiz & Stützle, 2007), denoted as IG_RS heuristic. IG method consists of two central procedures, destruction and construction. The initial sequence of IG_RS heuristic is generated by NEH heuristic. For destruction, IG_RS heuristic randomly removes a number of d jobs from the initial sequence resulting a partial sequence π_D; and for construction, IG_RS heuristic follows step 3 of NEH heuristic to insert each of d jobs back in to π_D. Heuristic development in phase 1 is beneficial for future heuristic development in the other two phases (Framinan et al., 2004).

Ruiz and Maroto (2005) compare 19 heuristics for $Fm/prmu/C_{max}$ problems, and concluded that NEH heuristic is the best, CDS heuristic the eighth, and two PDRs (LPT and SPT rules) the worst. However, CDS heuristic has the second simplest computational complexity among the first 8 heuristics, $O(m^2n+mnlogn)$. Moreover, King and Spachis (1980) compare 5 PDRs and CDS heuristic for two different TFS problems, $Fm/prmu/C_{max}$ and $Fm/nwt/C_{max}$. They conclude that CDS heuristic and LWBJD (least weighted between jobs delay) rule are the best for $Fm/prmu/C_{max}$ problems and MLSS (maximum left shift savings) rule is the best for $Fm/nwt/C_{max}$ problems, but no single method is consistently the best for both $Fm/prmu/C_{max}$ and $Fm/nwt/C_{max}$ problems.

The literature on HFS is still scarce (Linn & Zhang, 1999; Wang, 2005). According to Botta-Genoulaz (2000), CDS heuristic is the best of 6 heuristics for HFS problems, including NEH heuristic. The problem in Botta-Genoulaz (2000) is an n-job S-stage HFS problem to minimize the maximum lateness. It is converted to an n-job S+1-stage HFS problem to min(C_{max}). The processing time of job i in stage S+1 is calculated by $p_{i,S+1} = D_{max} - d_i$, $i = 1,\ldots,n$, where $D_{max} = \max(d_k)$, and d_k is the due date of job k, $k = 1,\ldots,n$. When applying CDS heuristic to HFS problems, Botta-Genoulaz converts the processing times, $p'_{i,j} = p_{i,j}/OPTR_j$, $j = 1,\ldots,S+1$, where $p_{i,j}$ is the original processing time of job i in stage j, and $OPTR_j$ is the number of operators/machines assigned to stage j.

For $FFs/nwt/C_{max}$ problems, Thornton and Hunsucker (2004) propose an NIS heuristic, the best among CDS heuristic, LPT and SPT rules, and a heuristic of random sequence generation. Different from CDS heuristic, NIS heuristic uses a filter concept to convert a $FFs/nwt/C_{max}$ problem to a virtual 2-machine problem, and then applies JA to get a job sequence. The stages before the filter are regarded as virtual machine 1, after the filter as virtual machine 2, and the stages that are covered by the filter are ignored. The filter goes from stage 2 to stage S–1, and the width of filter changes from 1 to S–2. In total, there are $1+(S-1)\times(S-2)/2$ sequences generated by NIS heuristic and the one with the minimum makespan is the final schedule.

2.2 Requirements for adaptive production control

2.2.1 Three criteria

Three main criteria are used to evaluate a heuristic for adaptive production scheduling and control (Li et al., 2011a): optimality, computational complexity, and flexibility. Usually optimality is used to evaluate a heuristic for offline production scheduling. However, when adaptive production control is taken into consideration, the computational complexity becomes critical. That is why some heuristics based on artificial intelligence are not suitable for adaptive production control, although they can get better solutions. Another criterion is the flexibility, that is, whether a heuristic can deal with a disturbance. Of course, different situations have different requirements for optimality, computational complexity, and flexibility of a heuristic. There is inevitably a trade-off among these criteria, and the selection of heuristics for production scheduling and control depends on specifics of different situations, such as the value of optimality as compared to near optimal scheduling, as well as the type and volume of disturbances that underlies the requirements of response time.

2.2.2 Summary of existing heuristics for adaptive production control

For optimality, heuristics in phase 3 can get better solutions than heuristics in phases 1 and 2. However, for computational complexity, they take much longer time. For example, an adaptive learning approach (ALA) heuristic is in phase 3 for $Fm/prmu/C_{max}$ problems (Agarwal et al., 2006). The deviation of ALA heuristic is only 1.74% for Taillard's benchmarks (Taillard, 1993), much better than 3.56% of NEH heuristic. However, for the largest instance in Taillard's benchmarks, i.e. 500 jobs and 20 machines, it takes more than 19 hours for ALA heuristic to get a solution, more than 20 hours for Simulated Annealing, and more than 30 hours for Tabu search (Agarwal et al., 2006). Even for the recent IG_RS heuristic, it takes 300 seconds to get a solution to a 500-job 20-machine instance. For flexibility, we need to see if a heuristic can deal with a disturbance. According to Pinedo (2002), there are three types of disturbances in general for flow shop production, job insertion or cancellation, operator absence or machine breakdown, and variation in processing times. The perfect production information in OKP is available only after the production (Wortmann, 1992). Therefore, if a heuristic operates the known processing time only, it cannot deal with variation in processing times.

The performance of first eight of 19 heuristics is summarized in Table 2.1, and the optimality of each heuristic is quoted from Ruiz and Maroto (2005). However, there is a discrepancy of optimality of heuristics in the literature, because the optimality is evaluated by the deviation from the best known upper bounds that are under continuous improvement. For example, the deviation of 3.33% is for NEH and 9.96% for CDS in Ruiz and Maroto (2005), but 3.56% for NEH and 10.22% for CDS in Agarwal et al. (2006), and 3.59% for NEH and 11.28% for CDS in our case study. In the table, the column "Opt." means the optimality on Taillard's benchmarks for $Fm/prmu/C_{max}$ problems, "I/C" means the job insertion or cancellation, "OA/MB" means the operator absence or machine breakdown, and "Var." means the variation in processing times. The mark of "Yes§" means a heuristic can deal with a disturbance only with a modification of processing times, e.g. in Botta-Genoulaz (2000).

		Computational Complexity		Flexibility		
	Opt.		Note	I/C	OA/MB	Var.
NEH	3.33%	$O(mn^2)$	$O(mn^3)$	Yes	Yes§	No
Suliman	6.21%	Intractable	CDS first, then swap job pairs	Yes	Yes§	No
RAES	7.43%	Intractable	RA first, then swap jobs	Yes	Yes§	No
HoCha	8.06%	Intractable	CDS first, then swap job pairs	Yes	Yes§	No
RACS	9.17%	Intractable	RA first, then swap jobs	Yes	Yes§	No
Koula	9.22%	$O(m^2n^2)$	JA first, then job passing	Yes	Yes§	No
HunRa	9.69%	$O(mn+n\log n)$	3 × Palmer's slope index	Yes	Yes§	No
CDS	9.96%	$O(m^2n+mn\log n)$	JA	Yes	Yes§	No

Table 2.1 Summary of 8 heuristics for adaptive production scheduling and control

It is self-illustrative for optimality and flexibility of each heuristic in the above table. We only discuss the computational complexity in the following. NEH heuristic, in its original version, has a computational complexity of $O(mn^3)$, but, by calculating the performance of all partial sequences in a single step, its complexity is reduced to $O(mn^2)$ (Taillard, 1990). Both Suliman (Suliman, 2000) and HoCha (Ho & Chang, 1991) heuristics use CDS heuristic to generate an initial sequence, and then exchange job pairs to improve the performance, but they use different mechanisms for job pair swaps. Because the number of job pair swaps depends on the calculation of performance of each job pair, the computational complexities of Suliman and HoCha heuristics are intractable. Job swaps are also involved in RACS and RAES heuristics (Dannenbring, 1977), and their computational complexities are intractable too. These two heuristics are based on a rapid access (RA) heuristic (Dannenbring, 1977), which is a mixture of JA and Palmer's slope index (Plamer, 1965). Koula heuristic (Koulamas, 1998) is not purely for permutation flow shop problems. The job passing is allowed in Koula heuristic, because Potts et al. (1991) point out that a permutation schedule is not necessarily optimal for all n-job m-machine flow shop problems. Koula heuristic extensively uses JA to generate initial sequences, and then job passing is allowed to make further improvement. The overall computational complexity of Koula heuristic is $O(m^2n^2)$. HunRa heuristic (Hundal & Rajgopal, 1988) is a simple extension of Palmer's slope index. HunRa heuristic generates three sequences, one by Palmer's slope index, the other two by calculating indices differently. Therefore, the HunRa heuristic has the same computational complexity as Palmer's slope index, $O(mn+n\log n)$. Usually, the number of jobs n is much larger than the number of machines m, thus, the computational complexity of $O(m^2n+mn\log n)$ for CDS heuristic is comparable with that of $O(mn+n\log n)$ for HunRa heuristic.

For an industrial instance in Gienow with 1396 jobs and 5 machines, it takes NEH heuristic more than 70 seconds to generate a sequence, which is too slow to keep up with the production pace in Gienow. Therefore, NEH and other five heuristics, with computational complexity higher than $O(mn^2)$, are not suitable for adaptive production scheduling and control in Gienow. It takes less than one second for CDS or HunRa heuristics to generate a sequence for the same industrial instance. However, their performance is not good from the optimality perspective, with more than 9% deviation on Taillard's benchmarks.

3. Adaptive production scheduling and control system

For adaptive production scheduling and control, it is necessary not only to monitor the production on the shop floor, but also to give a solution in time when a disturbance happens. Our computer-aided system for adaptive production scheduling and control in OKP consists of a close-loop structure and a state space (SS) heuristic.

3.1 The feedback control scheme

For adaptive production scheduling and control, a computer-aided scheduling and control system has been proposed as illustrated in Fig. 3.1, which consists of SS heuristic and a simulation model called temporized hierarchical object-oriented coloured Petri nets with changeable structure (THOCPN-CS) (Li, 2006). High customization and dynamic disturbances in OKP demand for a great effort on a simulation model. Simultaneously, adaptive production control demands for solutions in a short time. Therefore, the unique feature of the THOCPN-CS simulation model makes it easy and flexible to simulate frequent changes in OKP for adaptive production control. Steps to achieve adaptive production scheduling and control in OKP are summarized as follows.

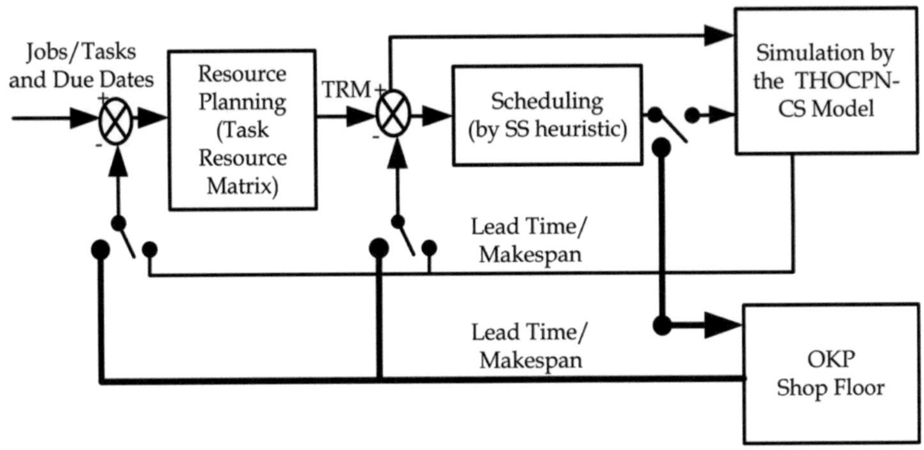

Fig. 3.1 A computer-aided production scheduling and control system

Step 1. Assign possible manufacturing resources (e.g. operators/machines) to each stage, and hence form a task-resource matrix (TRM) with jobs.

Step 2. Schedule the jobs by SS heuristic for offline scheduling, generating a sequence with the good performance for the next step.

Step 3. Simulate the production by the THOCPN-CS model, and identify the bottleneck stage(s) and overflow of WIP inventories. Human schedulers may carry out some adjustment to smooth the production flow, such as re-allocate operators/machines in stage(s), take some jobs away and then re-schedule the rest jobs, and so on.

Step 4. Re-schedule the jobs by both SS heuristic and human schedulers for offline scheduling. For online re-scheduling, re-schedule the jobs by either or both of the heuristic and scheduler, which depends on the time allowance for online re-scheduling.

Step 5. Repeat Steps 3 and 4 in the offline production scheduling phase until a satisfactory production schedule is obtained. This production schedule contains a job sequence and a number of operators/machines in each stage. In the adaptive production control phase, this step may be omitted, depending on specific requirements.

Step 6. Deliver the production schedule to the shop floor and switch the control loop from the simulation model to the shop floor.

Step 7. If any disturbance occurs on a shop floor, switch the control loop back to the simulation model, and go back to Step 3 if operators/machines re-allocation is necessary, or go back to Step 4.

Through repeating the above-mentioned steps iteratively, the production on OKP shop floors can be adaptively scheduled and controlled.

3.2 The state space heuristic

SS heuristic is mainly for HFS problems. Because there are multiple operators in each stage and the capacity of WIP inventories is limited, SS heuristic is not only to min(C_{max}), but also to maximize the utilization, max($Util$). There are two concepts used in SS heuristic, a state space concept and a lever concept.

3.2.1 The state space concept

Consider a hybrid flow line with 3 work stages and 2 operators in each stage (see Fig. 3.2).

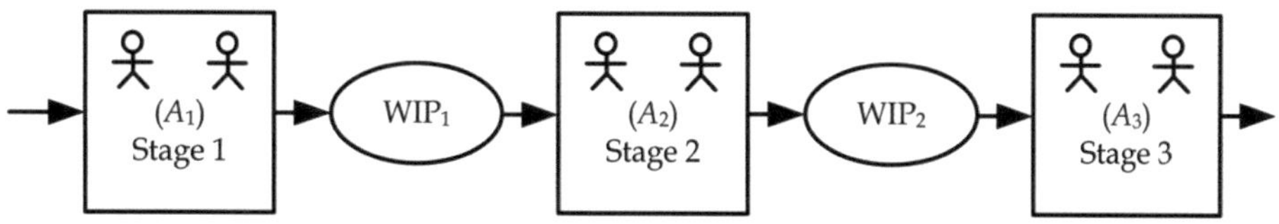

Fig. 3.2 A 3-stage flow line with 2 operators in each stage

The operators in each stage follow a *FCFS* rule. Then there is a next available time of each stage, A_s, where $A_s = \min(a_{s,k})$, for $k = 1,\ldots,OPTR_s$, in which $a_{s,k}$ is the next available time of operator k in stage s, and $OPTR_s$ is the number of operators in stage s. There are S–1 time differences between S-stage available times. In the example above, there are two differences of the next available times, A_2–A_1, and A_3–A_2. If we regard such a difference as a space, $SPACE_s = A_{s+1} - A_s$ for $s = 1,\ldots,S\text{-}1$, then $SPACE_s$ is a time period available for stage s to finish a job without causing idle to an operator in stage s+1. If the completion time of job i in stage s is larger than the next available time of stage s+1, then such a job causes idle to stage s+1, $IDLE_{i,s} = c_{i,s} - A_{s+1}$ where $c_{i,s}$ is the completion time of job i in stage s, $c_{i,s} = \max(A_s, c_{i,s\text{-}1}) + p_{i,s}$. If the completion time of job i in stage s is smaller than the next available time of stage s+1, then there are two possibilities depending on whether WIP is full. If the WIP inventory, WIP_s, is full, then a delay happens to operator k who processed job i in stage s, $DELAY_{i,s} = A_{s+1} - c_{is}$. Such a delay means that, after finishing job i, operator k in stage s has to hold it in hand for $DELAY_{i,s}$ time units until there is a vacancy in WIP_s. Therefore, the next available time of operator k in stage s is delayed. Alternatively, if WIP_s is not full, job i goes into the inventory, and there is no *IDLE* or *DELAY*.

The main idea of SS is to find a job that fits S–1 spaces, without causing *IDLE* or *DELAY*. After a job i is processed on a line, the next available times are changed, and the space is changed accordingly. Greater *IDLE* and *DELAY* are not good for production if objectives are to $\min(C_{max})$ and $\max(Util)$, while greater *SPACE* is good to some extent.

From the foregoing description of SS, we can see *IDLE* and *DELAY* are evaluated according to job i and stage s, but *SPACE* is only evaluated by stage s. To make *SPACE* both job and stage dependant, there are two ways to model *SPACE*. One model is $SPACE_{i,s} = c_{i,s+1} - A_s$, for $s = 1,\ldots,S\text{-}1$. The other model is $SPACE_{i,s} = p_{i,s+1}$, for $s = 1,\ldots,S\text{-}1$. In our current version of SS heuristic, we use the latter model of *SPACE*, reducing one calculation in iteration and increasing the computation speed for adaptive control. However, we illustrate the alternative model, $SPACE_{i,s} = c_{i,s+1} - A_s$, in section 4 to show the flexibility of SS concept.

3.2.2 The lever concept in SS

From our previous research on TFS problems, we find that the lever concept is suitable for flow shop production (Li et al., 2011b), which means *IDLE* (or *DELAY*) in an earlier stage is worse for $\min(C_{max})$ objective than in a later stage. Consider a lever where force F takes effect and causes a torque of $F \times L$, where F is the unit of force and L is the length of force arm. An S-stage flow line is modelled as a lever, and $IDLE_{i,s}$ or $DELAY_{i,s}$ has a torque effect manifested as $IDLE_{i,s} \times LVR\_IDLE_s$ or $DELAY_{i,s} \times LVR\_DELAY_s$.

The lever concept for *IDLE* in SS is shown in Fig. 3.3. For an *S*-stage flow line, a job could cause at most *S*–1 times of *IDLE*. No *IDLE* is caused in stage 1 and an *IDLE* takes effect in the next stage. Therefore, the fulcrum of a lever for *IDLE* is set between stages *S*–1 and *S*, and the length of arm for an *IDLE* caused by stage *s* in stage *s*+1 is $LVR\_IDLE_s = S-s$.

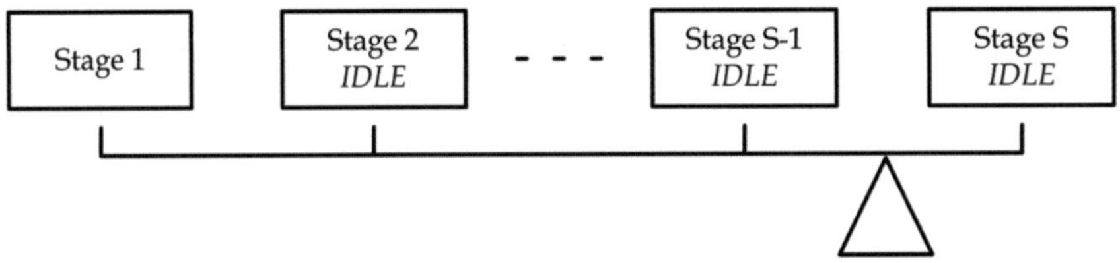

Fig. 3.3 A lever concept for *IDLE* in SS

The lever concept for *DELAY* in SS is shown in Fig. 3.4. Like the number of possible *IDLE*s, there could be *S*-1 *DELAY*s, and no *DELAY* in stage *S*. But a *DELAY* takes effect in current stage *s*, whereas *IDLE* is in the next stage. Therefore, one unit of *DELAY* in stage *s* should be worse than one unit of *IDLE* in stage *s*. Thus, the length of arm for a *DELAY* is $LVR\_DELAY_s = S-s+1$, for $s = 1,\ldots,S\text{-}1$. The fulcrum of a lever for *DELAY* is set in stage *S*.

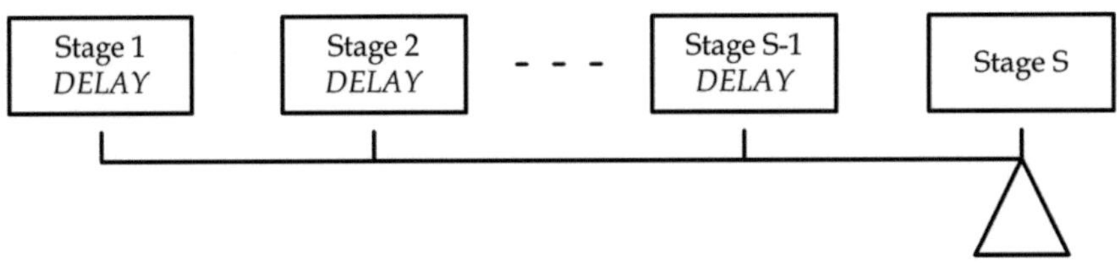

Fig. 3.4 A lever concept for *DELAY* in SS

There is also a lever concept for *SPACE* in SS, shown in Fig. 3.5. The length of force arm for a space is $LVR\_SPACE_s = s$, for $s = 1,\ldots,S\text{-}1$. The fulcrum of a lever for *SPACE* is set between stage 1 and stage 2, which means *SPACE* in a later stage is better than in an earlier stage.

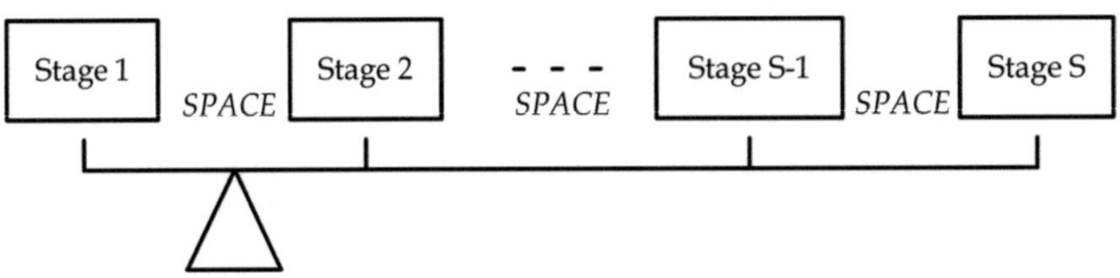

Fig. 3.5 A lever concept for *SPACE* in SS

Therefore, all *SPACE*s, *IDLE*s, and *DELAY*s are converted to torques, that is, $SPACE'_{i,s} = p_{i,s+1} \times LVR\_SPACE_s$, $IDLE'_{i,s} = IDLE_{i,s} \times LVR\_IDLE_s$, and $DELAY'_{i,s} = DELAY_{i,s} \times LVR\_DELAY_s$. The job selection scheme is $\max_{1\le i\le n}\left[\sum_{s=1}^{S-1} SPACE'_{i,s} - \left(\sum_{s=1}^{S-1} IDLE'_{i,s} + \sum_{s=1}^{S-1} DELAY'_{i,s}\right)\right]$, that is, to select the job with the maximum torque difference between *SPACE*'s and *IDLE*'s + *DELAY*'s.

3.2.3 Steps to achieve the SS heuristic

Two items should be taken into consideration for initial job selection in SS. One is the number of initial jobs, and the other is the initial job selection scheme. The number of initial jobs is set as min($OPTR_s$, for s = 1,...,S). The reason is that if the number of initial jobs is smaller than min($OPTR_s$), then the first available time of a stage is zero since all operators are available at time zero; if the number is greater than min($OPTR_s$), then the number of (initial job number - min($OPTR_s$)) jobs are not selected by the state space concept.

For initial job selection scheme, five 1×S vectors are introduced as follows: $Vector\_1 = [0]_{1\times S}$; $Vector\_3 = [APT_s]_{1\times S}$, where $APT_s = \sum_{i=1}^{N} p_{i,s} / n$ is the average processing time of stage s; $Vector\_5 = [\max(p_{i,s}), i = 1,\ldots,n]_{1\times S}$ for s =1,...,S is the maximum processing time of stage s; $Vector\_2 = Vector\_3/2$; $Vector\_4 = Vector\_3+[Vector\_5-Vector\_3]/2$. The initial number of jobs are selected according to $\min(\sum_{s=1}^{S} |p_{i,s} - Vector\_v(s)|)$ for i = 1,...,n, which means the minimum absolute difference between one job's processing times and the vector.

Step 1. Determine the number of operators in each stage, i.e. $OPTR_s$. (a): Calculate n and S. (b): Set an expected throughput rate, r, which means a job is to be finished in every r time units. (c): $OPTR_s$ = Roundup (APT_s/r). (d): Set the start time of every operator to 0. (e): Put all of n jobs into a candidate pool. (f): Set an output sequence to be a 1×n zero vector, *Sequence_v*.

Step 2. Set the capacity of each of S-1 WIP inventories.

Step 3. Calculate five vectors for initial job selection.

Step 4. FOR v = 1:5, an iteration loop to select initial jobs according to one *Vector_v*.

Step 5. Select a number of min($OPTR_s$, for s = 1,...,S) jobs according to *Vector_v* by the equation $\min(\sum_{s=1}^{S} |p_{i,s} - Vector\_v(s)|)$. Then put selected jobs into a *Sequence_v* and eliminate them from the candidate pool. Calculate the next available time of each operator, the next available time of each stage, namely *STATE*, and WIP inventory status, namely *WIP_Status*, which is initially a 1×(S–1) zero vector.

Step 6. FOR i = min($OPTR_s$) + 1: n, an iteration loop to sequence rest n - min($OPTR_s$) jobs.

Step 7. According to *STATE* and *WIP_Status*, calculate $IDLE'_{is}$, $DELAY'_{is}$ and $SPACE'_{is}$.

Step 8. Select job i according to $\max\limits_{1\le i\le n}\left[\sum_{s=1}^{S-1} SPACE'_{i,s} - \left(\sum_{s=1}^{S-1} IDLE'_{i,s} + \sum_{s=1}^{S-1} DELAY'_{i,s}\right)\right]$, and then put such job number into *Sequence_v* and eliminate it from the candidate pool.

Step 9. Calculate intermediate completion time of a partial schedule *Sequence_v*, update *WIP_Status*, and update *STATE*.

Step 10. END i. Calculate the utilization of a line. (a): Calculate utilization of each stage first, $Util_s = (\sum_{i=1}^{n} p_{i,s} / OPTR_s)/(c_{nks}-c_{1k's-1})$, $c_{1k0} = 0$, s = 1,...,S, in which c_{nks} is the completion time of the last job in stage s, and $c_{1k's-1}$ is the start time of the first job in stage s, i.e. the completion time of the first job in stage s-1. (b): Calculate the average utilization of each stage, i.e. $Util$ = average ($Util_s$), s = 1,...,S.

Step 11. ND v. Output each of five sequences and related makespan and utilization, and the minimum makespan and the maximum utilization are regarded as the final performance of SS.

3.2.4 The computational complexity of SS

The computational complexity of SS heuristic consists of two parts, job selection and makespan calculation.

For job selection, if the state of a flow line is known, then to select one out of n unscheduled jobs takes $S\times n$ operations, which means the computational complexity for adaptive control is $O(Sn)$. As n decreases from n to 1, the overall selection of n jobs takes $S\times n\times(n+1)/2$ operations. Although SS heuristic generates five sequences for an n by S HFS problem, the computational complexity of SS heuristic for job selection is $O(Sn^2)$, because only the highest order of operations is counted in computational complexity.

For makespan calculation, we can model an n-job S-stage HFS problem by a 2-dimension matrix, where the row dimension is for jobs, and the column dimension for stages. The makespan calculation could be carried out along the column dimension. It means that, if the input sequence of n jobs in stage 1 is known, then the output sequence of n jobs in stage 1 (or the input sequence in stage 2) can be calculated; the output sequence is in a non-descending order of completion times of n jobs; and then the output sequence in stage 2 can be calculated, and so on, finally the output sequence in stage S can be calculated. However, the capacities of WIP inventories are limited, which means the completion times of jobs in stage s are constrained by the next available times of operators in stage s+1. For example, when calculating the output sequence of stage s, if a job i's completion time in stage s causes an overflow of WIP_s, which means at that time the WIP_s is full and there is no operator available in stage s+1 to process a job in WIP_s, then a *DELAY* happens to such job i. This *DELAY* means the job i's completion time is delayed to a later time, and so is the next available time of operator k, who processes the job i in stage s. Consequently, the *DELAY* affects the completion times of all jobs following job i in stage s, and the completion times in the previous stage need to be checked because of the limitation on WIP inventories. In an extreme situation, when a *DELAY* happens in stage S-1, the job completion times in all previous stages have to be recalculated. Because of the recalculations, it is time consuming to calculate makespan along the column dimension.

For the makespan calculation along the row dimension, as n increases from 1 to n, the computational complexity is also $O(Sn^2)$, although makespan calculation is carried out five times. Therefore, the overall computational complexity of the SS heuristic is $O(Sn^2)$.

For an industrial instance with 1396 jobs and 5 machines in Gienow, the computation time of SS heuristic is 70.67 seconds, much longer than 782 milliseconds for CDS heuristic. However, the 782 milliseconds are only for CDS to generate sequences. Taking the makespan calculation into consideration, CDS will have the same computational complexity as SS. Moreover, from the adaptive control perspective, the computational complexity of SS heuristic is only $O(Sn)$, which means it takes only 10.12 milliseconds for SS heuristic to select the next job dealing with disturbances in Gienow for this instance.

4. Case studies

The computational complexities of some existing heuristics and SS heuristic are analyzed in sections 2 and 3 respectively. In this section, the comparison and evaluation of heuristics are mainly based on optimality and flexibility. Two kinds of case studies, with and without disturbances, are carried out on Taillard's benchmarks (Taillard, 1993) and on an industrial case. Section 4.1 is for without disturbances, Section 4.2 is for with disturbances, and at last Section 4.3 gives a comparison between SS heuristic and Johnson's algorithm (JA).

4.1 Case studies without disturbances

4.1.1 $Fm/prmu/C_{max}$ on Taillard's benchmarks

For traditional permutation flow shop scheduling problems, the deviation (*DEV*) from the best known upper bounds is used to evaluate the performance of a heuristic, where $DEV = (C_{max}\ \textit{of a heuristic} - \textit{The upper bound}) \div (\textit{The upper bound})$ in percentage. The results of the deviation studies for CDS, NIS and SS, and a version of SS without the lever concept, SSnoLVR, heuristics are shown in Table 4.1.

In Table 4.1, the column "Scale" means the size of problems. For example, 20*5 means 20-job 5-machine problems. The column "Inst" means the number of instances in each scale. Columns 3 to 6 represent the average deviation of each of the CDS, NIS, SS, and SSnoLVR, heuristics respectively. We can see that SS heuristic has the smallest total average deviation for all 120 instances in Taillard's benchmarks, at 8.11%, NIS heuristic ranks the second at 9.01%, and CDS heuristic ranks the last at 11.28%. We can also see from Table 4.1 that the lever concept is suitable for flow shop production to minimize the makespan. The SS heuristic is better than the SSnoLVR heuristic, with a deviation of 8.11% versus 8.80%. To further compare the performance of the SS heuristic with the CDS heuristic's, a *t*-test is carried out using a function of TTEST (CDS results, SS results, 2, 1) in excel. The the SS heuristic's *p*-value is 3.20×10^{-5}, which means the improvement is extremely significant.

Scale	Inst	CDS	NIS	SS	SSnoLVR
20*5	10	9.05	7.41	9.14	7.80
20*10	10	13.48	9.46	10.18	13.13
20*20	10	11.07	7.30	10.64	14.02
50*5	10	7.15	4.96	3.60	3.38
50*10	10	14.46	11.57	9.67	9.24
50*20	10	18.13	14.50	16.15	16.12
100*5	10	5.25	4.70	1.60	1.75
100*10	10	9.51	8.27	6.71	6.05
100*20	10	16.45	13.50	11.83	15.71
200*10	10	7.55	6.61	3.09	2.48
200*20	10	13.75	11.33	9.10	11.31
500*20	10	9.56	8.44	5.63	4.60
Total Average		11.28	9.01	8.11	8.80
MAX		21.13	16.62	20.83	22.02
MIN		0.66	0.86	0.78	0.60

Table 4.1 Average deviations from Taillard's benchmarks for $Fm/prmu/C_{max}$ problems (%).

4.1.2 $Fm/nwt/C_{max}$ on Taillard's benchmarks

For traditional no wait flow shop problems, an improvement (*IMPR*) over NIS heuristic is used to evaluate the performance of CDS and SS heuristics based on Taillard's benchmarks. $IMPR = (C_{max}$ of NIS $- C_{max}$ of CDS or SS$) \div (C_{max}$ of NIS$)$ in percentage is shown in Table 4.2.

Scale	Inst	CDS	SS
20*5	10	-0.32	2.01
20*10	10	-2.59	-2.86
20*20	10	-3.50	-2.71
50*5	10	0.29	8.29
50*10	10	-1.29	0.49
50*20	10	-2.42	-1.67
100*5	10	-0.27	9.20
100*10	10	-0.61	3.78
100*20	10	-1.00	-0.02
200*10	10	-0.22	5.59
200*20	10	-0.41	1.69
500*20	10	-0.10	3.46
Total Average		-1.04	2.27

Table 4.2 Improvement over NIS heuristic for *Fm/nwt/*C_{max} problems (%)

In Table 4.2, CDS heuristic performs 1.04% worse than NIS heuristic. In contrast, SS performs better than NIS on average, with an improvement of 2.27%. For the *t*-test based on 12 averages, SS heuristic's *p*-value is 0.0739. However, if the *t*-test is based on 120 individual cases, the SS' *p*-value is 2.07×10^{-11}, which means an extremely significant improvement. Moreover, we recognize that for HFS no wait problems the improvement of SS over NIS will shrink as the number of operators/machines in each stage increases. For example, if the number of operators in each stage is the same as the number of jobs, then C_{max} is fixed as max ($\sum_{s=1}^{S} p_{i,s}$) for $i = 1,\ldots,n$, no matter for no wait or no pre-emption flow shop problems.

4.1.3 *FFs/nwt/C* on Taillard's benchmarks

For hybrid no wait flow shop problems with identical parallel operators/machines in each stage, two operators/machines are assigned to each stage. The improvement of CDS and SS heuristics over NIS heuristic is shown in Table 4.3.

Scale	Inst	CDS	SS
20*5	10	-1.71	-2.66
20*10	10	-2.72	-2.02
20*20	10	-3.06	-2.88
50*5	10	-0.77	3.34
50*10	10	-1.50	-2.18
50*20	10	-3.48	-2.04
100*5	10	0.21	7.15
100*10	10	-0.55	0.60
100*20	10	-1.75	-1.13
200*10	10	-0.15	3.54
200*20	10	-0.50	0.97
500*20	10	0.01	2.00
Total Average		-1.33	0.39

Table 4.3 Improvement over NIS heuristic for *FFs/nwt/*C_{max} problems (%)

For such hybrid no wait flow shop problems with two operators/machines in each stage, SS heuristic has a small improvement of 0.39% over NIS heuristic, and CDS heuristic still performs worse, -1.33%. For the *t*-test, SS heuristic's *p*-value is 0.6739, meaning that its improvement over NIS heuristic is not statistically significant.

4.1.4 *FFs/FCFS/C_{max}* on Taillard's benchmarks

Scale	Inst	min(C_{max})	max(*Util*)
20*5	10	-2.39	7.33
20*10	10	0.27	5.66
20*20	10	-2.65	-0.02
50*5	10	2.87	4.90
50*10	10	2.47	6.17
50*20	10	0.08	1.45
100*5	10	2.42	3.47
100*10	10	1.34	4.69
100*20	10	1.54	2.43
200*10	10	3.14	3.96
200*20	10	2.03	4.41
500*20	10	2.79	3.14
Total Average		1.16	3.96

Table 4.4 Improvement over CDS heuristic for *FFs/FCFS/C_{max}* problems (%)

For HFS problems with the *FCFS* rule applied to jobs in WIP inventories, two variables are set. One is a throughput rate $r = 31$, used to calculate the number of operators in each stage, where $OPTR_s$ = Roundup (APT_s/r). The average processing time of each stage ranges from 30.75 to 64.40 for all of 120 instances in Taillard's benchmarks, therefore, $OPTR_s$ varies from 1 to 3 for each stage. Another variable is the capacity of WIP inventories. Different configurations of WIP inventories have different impacts on production (Vergara & Kim, 2009). For the ease of case study, the capacity of each WIP inventory is set the same, WIP_s = 5, even though in theory each could be set to a different value. The calculation of processing times in CDS is $p'_{i,s} = p_{i,s}/OPTR_s$, $s = 1,\ldots,S$ (Botta-Genoulaz, 2000). For the objective of min(C_{max}), the improvement (*IMPR*) of SS heuristic over CDS heuristic is used to evaluate performance, where $IMPR_1 = (C_{max}\ of\ CDS - C_{max}\ of\ SS) \div (C_{max}\ of\ CDS)$ in percentage. For the objective of max(*Util*), the improvement of SS heuristic over CDS heuristic is $IMPR_2 = (Util\ of\ SS - Util\ of\ CDS) \div (Util\ of\ CDS)$. The results are shown in Table 4.4.

For the objective of max(*Util*), SS heuristic has an average 3.96% improvement over CDS heuristic on Taillard's benchmarks, and for the objective of min(C_{max}), SS heuristic has an average 1.16% improvement. For the *t*-test, SS heuristic's *p*-value is 0.0666 for min(C_{max}) meaning the improvement over CDS heuristic is not quite statistically significant. However, for max(*Util*), the *p*-value of SS heuristic is 3.34×10^{-5}, an extremely significant improvement.

4.1.5 An industrial case study

To validate the SS heuristic in a real setting, an industrial case study was carried out in Gienow Windows and Doors, Canada. This case consists of 1396 jobs on a flow line with 5 stages for one-day production. These jobs are delivered to customers at a predetermined time in 28 batches. Each batch of products is destined for customers in a given geographic area. Using data provided by Gienow, SS heuristic produces the results shown in Table 4.5. In the SS heuristic, the *SPACE* is modelled as $SPACE_{i,s} = c_{i,s+1} - A_s$.

	Gienow	SS	IMPR		Gienow	SS	IMPR
1	1,795	1,711	84	16	1,489	1,489	0
2	1,458	1,444	14	17	1,477	1,477	0
3	1,698	1,697	1	18	1,743	1,712	31
4	2,292	2,261	31	19	1,751	1,745	6
5	1,570	1,556	14	20	1,434	1,430	4
6	1,798	1,753	45	21	1,587	1,570	17
7	1,420	1,420	0	22	1,587	1,393	194
8	1,573	1,567	6	23	1,196	1,165	31
9	1,828	1,805	23	24	1,094	1,083	11
10	1,676	1,676	0	25	1,362	1,362	0
11	1,568	1,568	0	26	1,281	1,281	0
12	1,691	1,691	0	27	923	923	0
13	1,465	1,465	0	28	857	851	6
14	1,364	1,353	11	Total	42,300	41,771	529
15	1,323	1,323	0	Percent			1.25%

Table 4.5 An industrial case study

As shown in Table 4.5, Gienow used 42,300 time units to finish 1396 jobs. The production of 1396 jobs in 42,300 time units was achieved by Gienow's original schedule, which was generated by an experienced production scheduler in Gienow. SS heuristic can generate new schedules, respectively reducing 42,300 time units to 41,771, a 1.25% improvement in productivity. Such improvement translates into the production of 17 additional products daily, or more than $5000 revenue per day. For the *t*-test, SS heuristic's *p*-value is 0.0164, which means the improvement is very significant.

4.2 Case studies with disturbances

To test the suitability of SS heuristic to adaptive production control, a case study of operator absence is carried out on Taillard's benchmarks. Modeling operator absence is the same as modeling machine breakdown. We assume that, when a half of jobs are finished, one operator is absent in the middle stage of a flow line, specifically in stage 3, 6, or 11 according to the scale of instances in Taillard's benchmarks. For the remaining half of the jobs, if the production is carried on according to the original schedule when such disturbances happen to the shop floor, then the completion time is recorded as *Original.* If adaptive control is applied, that is, using SS heuristic to re-schedule the remaining jobs, then the completion time is recorded as *Adaptive*. The improvement of adaptive control over no adaptive control is used to evaluate the performance, i.e. (*Original* – *Adaptive*) ÷ (*Original*) in percentage.

To show the potential of the SS heuristic, case studies on operator absence are carried out under the two definitions of *SPACE*, $SPACE_{i,s} = p_{i,s+1}$ and $SPACE_{i,s} = c_{i,s+1} - A_s$. Moreover, a simple optimization method is also integrated with the SS heuristic.

4.2.1 $SPACE_{i,s} = p_{i,s+1}$

The results are given in Table 4.6. As we see, adaptive control is slightly better than no adaptive control with a 0.10% improvement for the SS heuristic if $SPACE_{i,s} = p_{i,s+1}$.

Scale	Inst	SS
20*5	10	2.46
20*10	10	1.81
20*20	10	3.01
50*5	10	0.88
50*10	10	2.17
50*20	10	-2.80
100*5	10	0.39
100*10	10	0.29
100*20	10	-4.18
200*10	10	-0.41
200*20	10	-1.25
500*20	10	-1.21
Total Average		0.10

Table 4.6 Adaptive control over no adaptive control, where $SPACE_{i,s} = p_{i,s+1}$

4.2.2 $SPACE_{i,s} = c_{i,s+1} - A_s$

The results are given in Table 4.7. As we see, for SS heuristic, if we model $SPACE_{i,s} = c_{i,s+1} - A_s$, the adaptive control has a 2.02% improvement over no adaptive control.

Scale	Inst	SS
20*5	10	7.75
20*10	10	6.62
20*20	10	-8.49
50*5	10	1.23
50*10	10	3.10
50*20	10	2.99
100*5	10	0.22
100*10	10	3.26
100*20	10	4.50
200*10	10	0.55
200*20	10	1.64
500*20	10	0.86
Total Average		2.02

Table 4.7 Adaptive control over no adaptive control, where $SPACE_{i,s} = c_{i,s+1} - A_s$

4.2.3 Integration with an optimization method

There are two effects in SS heuristic impacting the final production performance. *SPACE* is good for production but "*IDLE & DELAY*" is bad. We can introduce a weighting factor, α, into SS heuristic, and then sequence jobs according to max[(1–α)× $\sum_{s=1}^{S-1} SPACE'_{is}$ – α×($\sum_{s=1}^{S-1} IDLE'_{is}$ + $\sum_{s=1}^{S-1} DELAY'_{is}$)]. As α changes from 0 to 1 with increments of 0.1, the performance of SS heuristic, with $SPACE_{i,s}$ = $p_{i,s+1}$, is shown in Table 4.8. The columns represent the performance of each α integrated with SS heuristic. A weight α = 0.0 means no *IDLE* or *DELAY* is taken into consideration to sequence jobs, and α = 1.0 means no *SPACE*. We can see that *SPACE* affects the production more than *IDLE* or *DELAY*, where α = 0.1 has the greatest improvement of 2.77%.

Scale	Inst	0.0	0.1	0.2	0.3	0.4	0.5	0.6	0.7	0.8	0.9	1.0
20*5	10	-0.59	0.75	2.17	3.12	1.97	2.46	-2.31	-7.08	-7.68	-9.25	-2.30
20*10	10	0.67	0.77	0.56	5.01	3.68	1.81	-2.86	-4.62	-2.11	-1.59	-4.27
20*20	10	7.50	10.0	8.69	6.74	6.15	3.01	0.00	-9.46	-9.12	-8.99	-9.79
50*5	10	0.22	1.11	1.17	0.97	1.55	0.88	0.15	-1.66	-1.33	-1.24	-0.64
50*10	10	5.50	4.97	4.06	2.80	2.75	2.17	-4.35	-7.10	-6.20	-8.76	-8.89
50*20	10	1.65	3.67	1.94	2.90	-5.41	-2.80	-5.53	-5.22	-7.50	-0.22	-1.07
100*5	10	0.43	0.88	0.36	0.49	0.58	0.39	0.17	-0.98	-1.09	-1.16	-2.24
100*10	10	3.05	3.02	2.20	2.42	1.05	0.29	-2.35	-1.97	-2.76	-1.42	-2.50
100*20	10	5.94	5.24	5.79	4.76	-0.39	-4.18	-5.26	-5.76	-6.61	-5.73	-7.49
200*10	10	0.47	0.47	0.05	0.27	0.15	-0.41	-0.75	-1.26	-1.10	-1.58	-1.15
200*20	10	1.53	2.24	2.39	0.82	-0.15	-1.25	-1.44	-1.48	-2.34	-2.36	-1.98
500*20	10	0.17	0.18	0.16	-0.02	-0.64	-1.21	-1.63	-1.69	-1.66	-1.59	-1.54
Total average		2.21	2.77	2.46	2.52	0.94	0.10	-2.18	-4.02	-4.96	-5.33	-6.15

Table 4.8 Adaptive control with α over no adaptive control, where $SPACE_{i,s}$ = $p_{i,s+1}$

4.2.4 Case studies on variation in processing times

It is normal to have variation in processing times, especially for the production of highly customized products and with manual operations. Thus, it is necessary to test the suitability of SS heuristic to the disturbance of variation in processing times. In Gienow, processing times of products are quoted by the company standards.

For variation in processing times in the industrial case, we assume that, initially, we have a matrix of quoted processing times of *n* jobs in *S* stages, and we do not know the real processing time beforehand, because the perfect production information in OKP can be available only after the production (Wortmann, 1992). If we define this matrix as *B*, which means before production, we carry out the offline scheduling according to *B* to get a sequence *SB*. We might setup due dates based on the performance of *SB*, that is, the original performance, *PO*. After the actual production, we have a matrix of real processing times of *n* jobs in *S* stages, i.e. matrix *A*.

During the production, when variation in processing times happens and the production is carried out according to the sequence *SB*, the performance is *PB*. It means no adaptive control. It is difficult to use CDS heuristic for such disturbance, because we only know part of matrix *A* for finished jobs, but not the rest for unfinished jobs. However, we can adaptively re-schedule the rest jobs by SS heuristic, because the actual processing times of finished jobs affect the space, although we only know matrix *B* for the unfinished jobs. After one job has been produced, we use SS heuristic to select a job from remaining jobs according to processing times of unfinished jobs in matrix *B* and the actual space created by finished jobs. Consequently, the performance of adaptive control by SS heuristic is *PA*. The *SPACE* of SS heuristic is modeled as $SPACE_{i,s} = c_{i,s+1} - A_s$. We compare the performance of no adaptive control *PB* or adaptive control *PA* with the original performance *PO*, by: *Diff_OB* = (*PB* - *PO*) ÷ *PO* and *Diff_OA* = (*PA* - *PO*) ÷ *PO*, both of which are in percentage. Four ranges of normal distribution are introduced into the processing times in the industrial case, [-5%, 5%], [0%, 10%], [0%, 25%] and [0%, 50%]. The results are summarized in Tables 4.9.

		Diff_OB	*Diff_OA*
[-5%, 5%]	Average	0.83	0.27
	MAX	3.76	1.17
	MIN	0.01	0.03
[0%, 10%]	Average	5.45	5.12
	MAX	7.89	5.88
	MIN	4.00	4.47
[0%, 25%]	Average	13.08	12.76
	MAX	15.72	15.69
	MIN	9.72	9.37
[0%, 50%]	Average	25.37%	24.98
	MAX	28.75%	28.19
	MIN	20.94%	20.38

Table 4.9 Adaptive control for variation in processing times

From Table 4.9, we can see that adaptive control performs better than no adaptive control for all four ranges of variation in processing times. Moreover, SS heuristic is stable to such disturbance, because its average difference of performance is close to the expected value of variation in four ranges respectively.

4.3 A case study on a 2-machine flow shop problem

To reveal the rationale and coherent logic of the state space concept, a scaled down version of SS heuristic is compared with JA for a 2-machine flow shop problem, $F2/prmu/C_{max}$. For $F2/prmu/C_{max}$ problems, the lever concept has no effect on the job selection in SS heuristic. This is because for this type of $F2/prmu/C_{max}$ problems, the WIP inventory between machines 1 and 2 is unlimited, thus no *DELAY* is taken into consideration. In addition, the length of force arm for *SPACE* or *IDLE* equals to one. However, the state space concept can yield different job sequences than JA. A numerical example is provided in Table 4.10.

	M_1	M_2
Job 1	5	20
Job 2	20	10
Job 3	10	15
Job 4	15	12

Table 4.10 A 2-machine flow shop example

JA sequences jobs according to the following scheme. If min $\{p_{i,1}, p_{i+1,2}\} \leq$ min $\{p_{i+1,1}, p_{i,2}\}$, then job i should be processed earlier than job i+1. Therefore, for the example in the above table,

JA generates a sequence of [Job 1, 3, 4, 2] with C_{max} = 62. According to the state space concept (but not exactly SS heuristic), and using JA for the initial job selection, two additional sequences can be obtained, [Job 1, 2, 3, 4] and [Job 1, 4, 3, 2], both of which have C_{max} = 62, and are different from the one generated by JA. Therefore, it is obvious that JA uses a sufficient condition for $F2/prmu/C_{max}$ problems, but not necessary in some cases. The state space concept can yield different sequences than JA with the same level of optimality, and hence can provide greater opportunities for improvement as the core of a more elaborate heuristic for adaptive production scheduling and control.

5. Conclusions and future work

One-of-a-kind production (OKP) challenges production scheduling differently from mass production, because of high customer involvement in OKP. Especially, it challenges production control severely, because of dynamic disturbances. Traditionally, offline production scheduling is separated from the online adaptive production control. Dynamic disturbances in OKP fail the production schedule, which are generated by heuristics that are developed based on strong assumptions for offline scheduling (MacCarty & Liu, 1993). Accordingly, adaptive production control is in need to deal with disturbances. Currently, the adaptive production control in OKP companies is carried out by shop floor managers using priority dispatching rules (PDRs) and their experience. However, the performance of PDRs is poor on most scheduling objectives (Ruiz & Maroto, 2005), and the experience might be good for local optimization but definitely lacks global optimization for the overall production. Therefore, the adaptive production scheduling and control is essential and indispensable to improve the production efficiency in OKP.

In regards to three criteria of optimality, computational complexity, and flexibility to evaluate a heuristic for adaptive production control (Li et al., 2011a), the state space (SS) heuristic is the better than most existing heuristics. For optimality, SS heuristic outperforms the most popular alternative heuristics (CDS, NIS) against Taillard's benchmarks no matter for $Fm/prmu/C_{max}$, $Fm/nwt/C_{max}$, $FFs/nwt/C_{max}$ and $FFs/FCFS/C_{max}$ problems. In addition, the production schedule generated by SS heuristic outperforms Gienow's original schedule, improving Gienow's daily productivity by 1.25%. For computational complexity, $O(m^2n+mnlogn)$ of CDS heuristic is simpler one than $O(mn^2)$ of SS heuristic for offline scheduling. However, if taking sequence evaluation into consideration, they have the same computational complexity of $O(mn^2)$. In addition, for online adaptive production control, the computational complexity of SS decreases to $O(mn)$, but that of CDS keeps the same. For flexibility, SS heuristic is more flexible than other heuristics. SS heuristic can deal with all three typical disturbances proposed by Pinedo (2002), job insertion or cancellation, operator absence or machine breakdown, and variation in processing times, whereas, CDS cannot deal with variation in processing times. Although NEH heuristic has the best performance for $Fm/prmu/C_{max}$ problems, its inflexible procedure to construct a job sequence renders it little flexibility to deal with disturbances. Moreover, SS heuristic is in the phase of index development, a phase that is beneficial for heuristic development in the other two phases (Framinan et al. 2004).

As discussed in this chapter, adaptive production scheduling and control in OKP challenges nearly all existing scheduling algorithms and heuristics, and almost all manufacturing companies are facing a certain level of disturbances, such as unreliable supplies, unexpected operator absence, machine breakdowns, etc. There is still a gap between the theoretical research and industrial applications. Industrial applications require further understandings and studies of production scheduling and control. This draws the following future work. (1) Production planning on the company level should be integrated with production scheduling and control on the shop floor level. Production planning provides a company the production capacity that is a constraint for adaptive production scheduling and control. Meanwhile, the adaptive production scheduling and control requires frequent re-planning according to the production progress under unexpected disturbances. This is to meet due dates of customer orders or provide better estimated lead-times. The synergy and co-optimization between these two levels are necessary and should be further researched. (2) Consequently, adaptive production scheduling and control for non-deterministic problems is inevitable. Stochastic modeling or simulation for non-deterministic production problems is a valuable research topic and lucrative. (3) It is critical to integrate material flows on shop floors into a supply chain to successfully achieve adaptive production scheduling and control in OKP. This is in fact an urgent research topic to be studied.

6. References

Agarwal, A.; Colak, S. & Eryarsoy, E. (2006). Improvement heuristic for the flow-shop scheduling problem: An adaptive-learning approach. *European Journal of Operational Research*, Vol.169, No.3, pp. 801-815

Blecker, T. & Friedrich, G. (Eds.) (2006). *Mass Customization: Challenges and Solutions*. Springer, ISBN 987-038-7322-22-3, New York, USA

Botta-Genoulaz, V. (2000). Hybrid flow shop scheduling with precedence constraints and time lags to minimize maximum lateness. *International Journal of Production Economics*, Vol.64, No.1, pp. 101-111

Campbell, H.G.; Dudek, R.A. & Smith, M.L. (1970). A heuristic algorithm for the n-job, m-machine scheduling problem. *Management Science*, Vol.16, No.10, pp. 630-637

Dannenbring, D.G. (1977). An evaluation of flow shop sequencing heuristics. *Management Science*, Vol.23, No.11, 1174–1182

Dean, P.R.; Tu, Y.L. & Xue, D. (2009). An Information System for One-of-a-Kind Production. *International Journal of Production Research*, Vol.47, No.4, pp. 1071-1087

Dean, P.R.; Tu, Y.L. & Xue, D. (2008). A Framework for Generating Product Production Information for Mass Customization. *International Journal of Advanced Manufacturing Technology*, Vol.38, No.11-12, pp. 1244-1259

Dudek, R.A. & Teuton Jr., O.F. (1964). Development of *M*-state decision rule for scheduling *n* jobs through *M* machines. *Operations Research*, Vol.12, No.3, pp. 471-497

Framinan, J.M.; Gupta, J.N.D. & Leisten, R. (2004). A review and classification of heuristics for permutation flow-shop scheduling with makespan objective. *Journal of the Operational Research Society*, Vol.55, No.12, pp. 1243-1255

Garey, M.R.; Johnson, D.S. & Sethi, R. (1976). The complexity of flowshop and jobshop scheduling. *Mathematics of Operations Research*, Vol.1, No.2, pp. 117-129

Goyal, S.K.; Mehta, K.; Kodali, R. & Deshmukh, S.G. (1995). Simulation for analysis of scheduling rules for a flexible manufacturing system. *Integrated Manufacturing Systems*, Vol.6, No.5, pp. 21-26

Graham, R.L.; Lawler, E.L.; Lenstra, J.K & Rinnooy Kan, A.H.G. (1979). Optimization and approximation in deterministic sequencing and scheduling: A survey. *Annals of Discrete Mathematics*, Vol.5, pp. 287-326

Gupta, J.N.D. & Stafford, E.F. (2006). Flowshop Research after Five Decades. *European Journal of Operational Research*, Vol.169, No.3, pp 699-711

Ho, J.C. & Chang, Y.L. (1991). A new heuristic for the n-job, m-machine flow-shop problem. *European Journal of Operational Research*, Vol.52, pp. 194–202

Hundal, T.S. & Rajgopal, J. (1988). An extension of Palmer's heuristic for the flow shop scheduling problem. *International Journal of Production Research*, Vol.26, No.6, pp. 1119-1124

Ignall, E. & Schrage, L. (1965). Application of branch-and-bound technique to some flow shop problems. *Operations Research*, Vol.13, No.3, pp. 400-412

Johnson, S.M. (1954). Optimal two- and three-stage production schedules with setup times included. *Naval Research Logistics Quarterly*, Vol.1, No.1, pp. 61-68

Karush, W. (1965). A counterexample of a proposed algorithm for optimal sequencing of jobs. *Operations Research*, Vol.13, No.2, pp. 323-325

King, J.R. & Spachis, A.S. (1980). Heuristics for flow-shop scheduling. I*nternational Journal of Production Research*, Vol.18, No.3, pp. 345-357

Koulamas, C. (1998). A new constructive heuristic for the flowshop scheduling problem. *European Journal of Operational Research*, Vol.105, No.1, pp. 66-71

Kouvelis, P.; Chambers, C. & Yu, D.Z. (2005). Manufacturing operations manuscripts published in the first 52 issues of POM: review, trends, and opportunities. *Production and Operations Management*, Vol.14, No.4, pp. 450-467

Lageweg, B.J.; Lenstra, J.K. & Rinnooy Kan, A.H.G. (1978). A general bounding scheme for the permutation flow-shop problem. *Operations Research*, Vol.26, No.1, pp. 53-67

Li, W. (2006). Adaptive *Production Scheduling and Control in One-of-a-Kind Production*, Thesis (M.Sc.), University of Calgary, Canada

Li, W.; Nault, B.R.; Xue, D. & Tu, Y.L. (2011a). An efficient heuristic for adaptive production scheduling and control in one-of-a-kind production. *Computers & Operations Research*, Vol.38, No.1, pp. 267-276

Li, W.; Luo, X.G.; Xue, D. & Tu, Y.L. (2011b). A heuristic for adaptive production scheduling and control in flow shop production. *International Journal of Production Research*, Vol.49, No.11, pp. 3151–3170

Linn, R. & Zhang, W. (1999). Hybrid flow shop scheduling: a survey. *Computers & Industrial Engineering*, Vol.37, No.1, pp. 57-61

MacCarthy, B.L. & Liu, J. (1993). Addressing the gap in scheduling research: a review of optimization and heuristic methods in production scheduling. *International Journal of Production Research*, Vol.31, No.1, pp. 59-79

McKay, K.; Pinedo, M. & Webster, S. (2002). Practice-focused research issues for scheduling systems. *Production and Operations Management*, Vol.11, No.2, pp. 249-258

Nawaz, M.; Enscore, E.E. & Ham, I. (1983). A heuristic algorithm for the m-machine, n-job flow-shop sequencing problem. *OMEGA The International Journal of Management Science*, Vol.11, No.1, pp. 91-95

Ovacik, I.M. & Uzsoy, R. (1997). *Decomposition Methods for Complex Factory Scheduling Problems*. Kluwer Academic Publishers, ISBN 079-2398-351, Boston, USA

Palmer, D. (1965). Sequencing jobs through a multi-stage process in the minimum total time - a quick method of obtaining a near optimum. *Operational Research Quarterly*, Vol.16, No.1, pp. 101-107

Park, S.C.; Raman, N. & Shaw, M.J. (1997). Adaptive scheduling in dynamic flexible manufacturing systems: a dynamic rule selection approach. *IEEE Transactions on Robotics and Automation*, Vol.13, No.4, pp. 486-502

Pinedo, M. (2002). *Scheduling Theory, Algorithms, and Systems*. Prentice Hall, ISBN 013-0281-387, New Jersey, USA

Potts, C.N.; Shmoys, D.B. & Williamson, D.P. (1991). Permutation vs. non-permutation flow shop schedules. *Operations Research Letters*, Vol.10, No.5, pp. 281-284

Ruiz, R. & Maroto, C. (2005). A comprehensive review and evaluation of permutation flowshop heuristics. *European Journal of Operational Research*, 165, No.2, pp. 479-494

Ruiz, R. & Stützle, T. (2007). A simple and effective iterated greedy algorithm for the permutation flowshop scheduling problem. *European Journal of Operational Research*, Vol.177, No.3, pp. 2033-2049

Smith, R.D. & Dudek, R.A. (1967). A general algorithm for solution of the n-job M-machine sequencing problems of the flow shop. *Operations Research*, Vol.15, No.1, pp. 71-82 Also see their correction (1969). Errata. *Operations Research*, Vol.17, No.4, pp. 756

Suliman, S. (2000). A two-phase heuristic approach to the permutation flow-shop scheduling problem. *International Journal of Production Economics*, Vol.64, No.1-3, pp. 143-152

Szwarc, W. (1971a). Elimination methods in the $m \times n$ sequencing problem. *Naval Research Logistics Quarterly*, Vol.18, No.3, pp. 295-305

Szwarc, W. (1971b). Optimal elimination methods in the $m \times n$ flow-shop scheduling problem. *Operations Research*. Vol.16, No.3, pp. 250-1259

Taillard E. (1993). Benchmarks for basic scheduling problems. *European Journal of Operational Research*, Vol.64, No.2, pp. 278-285

Taillard, E. (1990). Some efficient heuristic methods for the flow-shop sequencing problem. *European Journal of Operational Research*, Vol.47, No.1, pp. 65-74

Thornton, H.W. & Hunsucker, J.L. (2004). A new heuristic for minimal makespan in flow shops with multiple processors and no intermediate storage. *European Journal of Operational Research*, Vol.152, No.1, pp. 96-114

Tu, Y.L. (1996a). A Framework for Production Planning and Control in a Virtual OKP Company. *Trans. North American Manufacturing Research Institution of SME*, Vol.24, pp. 121-126

Tu, Y.L. (1996b). Automatic Scheduling and Control of a Ship Welding Assembly Line. *Computers in Industry*, Vol.29, No.3, pp. 169-177

Vergara H.A. & Kim, D.S. (2009). A new method for the placement of buffers in serial production lines. *International Journal of Production Research*, Vol.47, No.16, pp. 4437-4456

Wang, H. (2005). Flexible flow shop scheduling: optimum, heuristic and artificial intelligence solutions. *Expert Systems*, Vol.22, No.2, pp. 78-85

Wortmann, J.C. (1992). Production management systems for one-of-a-kind products. *Computers in Industry*, Vol.19, No.1, pp. 79-88

Wortmann, J.C.; Muntslag, D.R. & Timmermans, P.J.M. (1997). *Customer-Driven Manufacturing*. Chapman & Hall, ISBN 041-2570-300, London, UK

Part 4

Modeling and Solving Scheduling Problems in Practice

7

Simulation-Based Modular Scheduling System of Semiconductor Manufacturing

Li Li, Qiao Fei Ma Yumin and Ye Kai
Tongji University,
China

1. Introduction

Discrete event simulation technology, imitating a real production environment by modelling, has been applied to the scheduling of semiconductor manufacturing fabrication facilities (fabs) since 1980s.

For example, Wein (1988) compared various release control strategies and scheduling rules and provided the better combinations of release control strategy and scheduling rule for three kinds of HP 24 models. Peyrol et al. (1993) implemented a simulated annealing methodology to determine the input order of a given set of products so as to minimize the average residence time of these products in the plant with unlimited intermediate storage policy. They developed a discrete simulation program to determine the value of the objective function. This method was applied to a semiconductor circuit fabrication plant of MOTOROLA Inc. Thompson (1996) described a simulation-based finite capacity planning and scheduling software to allow human planners to make better decisions. This tool improved cycle time, throughput, and equipment utilization significantly without adding additional equipment and personnel. Baudouin et al. (1995) presented information system architecture and a decision support tool to allow admissible scheduling solutions in a semiconductor device manufacturing fab.

Presently, discrete event simulation technology has been widely applied to predict the operational performance of a semiconductor manufacturing fabrication facility (fab) and assist its scheduling decisions. Existing fruitful research results can be classified into three main directions.

Simulation systems performing scheduling behaviors

Some researchers developed various simulation systems to perform or assist scheduling behaviours of semiconductor wafer fabs.

Ramírez-Hernández et al. (2010a, 2010b) presented an architecture and implementation of a preventive maintenance optimization software tool, based on Approximate Dynamic Programming algorithms, for the optimal scheduling of preventive maintenance tasks in semiconductor manufacturing operations.

Weigert et al. (2009) developed a simulation-based scheduling system for a semiconductor Backend facility. The heuristic search strategies were adopted to optimize the operating sequences with consideration on concurrent minimization of mean cycle time, maximization of throughput and due date compliance.

Horn et al. (2006) developed a simulation-based interactive scheduling system, called "BackendPlanner", for the backend facility of a semiconductor manufacturer under practical conditions. Its special characteristics included the automated model generation from existent production databases, a heuristic optimization component, a fast-simulator, and the efficient data coupling strategy allowed the re-scheduling anytime, e.g., after machine breakdowns or other unpredictable events and several additional analysis tools reported important parameters of the manufacturing process, such as machine utilization, throughput time or work in process.

Zhang et al. (2006, 2009) implemented a simulation platform based on extended object-oriented petri net for real-time scheduling of semiconductor wafer fabrication systems, and designed a dynamic bottleneck dispatching algorithm to detect bottlenecks in a timely way to make adaptive dispatching decisions according to the real-time conditions.

In view of semiconductor probe system's production scheduling problem, Zhang and Wang (2008) set up a probe behaviour model in terms of the figuration of UML and transformed it directly into simulation model by using the SimTalk of eM-plant. The experimental result of simulation model supplied the decision maker with evidence.

Werner et al. (2006) developed a scheduling system for the backend of Infineon Technologies Dresden based on a discrete event simulation system and tested it in the real industrial environment. The simulation model was automatically generated from the databases of the manufacturer and was used for short term scheduling - from one shift up to one week.

Sivakumar and Gupta (2006) conceptualized, designed, and developed a discrete event simulation based "online near-real-time" dynamic multi-objective scheduling system to achieve Pareto optimal solutions in a complex manufacturing environment of semiconductor back-end. The system used a linear weighted aggregation optimization approach for multiple objectives and auto simulation model generation for online simulation. In addition, it enabled managers and senior planners to carry out "what now" analysis to make effective current decisions and "what if" analysis to plan for the future.

Simulation as a tool to verify and improve the performance of scheduling algorithms

Comparing to the achievements on the simulation-based scheduling system, the results of simulation as a tool to verify and improve the performance of scheduling algorithms are much more comprehensive.

Jeng and Tsai (2010) applied simulation experiments to show their Match Due Date scheduling rule focused on the actual working mode of memory IC with the ability to reduce earliness and tardiness, make less number of setups and shorten the flow time and enhance the confirmed line item performance.

Liu et al. (2010) presented dynamic scheduling models based on the resource conflict resolution strategies for a semiconductor production system and validated the model with a simulation case.

Tang et al. (2010) proposed a genetic algorithm and simulated annealing algorithm based scheduling method in semiconductor manufacturing lines and verified it with a Minifab simulation model.

Chen and Wang (2009) proposed a modified fluctuation smoothing rule incorporating a fuzzy-neural remaining cycle time estimator to improve scheduling performance in a semiconductor manufacturing factory, and evaluated its effectiveness by production simulation.

Zhang et al. (2007) combined different dispatching rules and rework strategies to obtain better performance of cycle time, work in process, on-time delivery, and throughput. The simulation results showed the superiority of the proposed scheduling to static scheduling methods.

Shi et al. (2008) used the simulation to determine the parameter values of a heuristic rule of a semiconductor manufacturing system and concluded a rule for the parameters' selection.

Chou et al. (2008) developed a simulated annealing algorithm with a probability matrix integrated with a greedy heuristic to solve the dynamic scheduling problem of semiconductor burn-in operations optimally in practical sizes. They proved that the proposed method could effectively and efficiently obtain optimal solutions for small size of problems and provide high-quality solutions efficiently for large size of problems by simulating.

The evaluation and improvement of simulation-based scheduling methods

The importance of simulation-based scheduling methods has aroused many researchers' interests. They wanted to improve this method better than ever before.

For example, Li and Mason (2007) investigated the potential advantages and drawbacks of using simulation-based scheduling in a semiconductor wafer fab. Results suggested the potential for simulation-based scheduling approaches to improve on-time delivery performance to customers in wafer fabs.

Koyuncu et al. (2007) augmented the validity of simulation models in the most economical way via incorporating dynamic data into the executing model, which then steered the measurement process for selective data update.

Kim et al. (2003) proposed a simplification method for accelerating simulation-based real-time scheduling in a semiconductor wafer fabrication facility. In the suggested real-time scheduling method, lot scheduling rules and batch scheduling rules were selected from sets of candidate rules based on information obtained from discrete event simulation. Since a rule combination that gave the best performance may vary according to the states of the fab, they suggested three techniques for accelerating rule comparison, too.

Although the simulation-based scheduling approaches in semiconductor manufacturing field have made big progresses, their common problem is lack of a standard scheduling simulation model. Consequently, quite a few of research results are applicable to a special production environment and difficult to be directly extended to other semiconductor wafer fabs to obtain better performance. Presently, there are some exploratory researches on standard simulation model. For instance, Ralph et al. (2007) presented a prototype framework for scheduling of semiconductor manufacturing that divided the production system into Fabmodel, Process Route, Process Step, Tool Set and Operator Set. Rajesh and Sen (2004) proposed a method to simplify the simulation models' complexity.

In this paper, we design a standard simulation-based modular scheduling system (SMSS) by means of modularization and decoupling the algorithms from the models according to scheduling characteristics of semiconductor wafer fabs. The remainder of this book chapter is organized as follows. A general structure model (GSM) of the scheduling of semiconductor wafer fabs is introduced in section 2. Section 3 presents a data-based dynamic simulation modeling method. Section 4 designs and develops a Simulation-based Modular Scheduling System (SMSS). Finally, we validate SMSS with a case study in section 5. Section 6 contains conclusions and future works.

2. General structure model

In view of its physical structure, a semiconductor wafer fab is composed of dozens of work-centres. Each work-centre includes multiple machines. Every machine has several or dozens of recipes to finish more than one operation of a job. An operation of a job can be finished by one or more machines and its processing time on different machines may be different.

In view of its process flow, a job will be processed as soon as it is released to a wafer fab. The machine processes a job according to an existing sequencing plan or directly selects a job in its queue according to some priority rule considering the real states of the wafer fab. A job's processing record is stored in the manufacturing execution system (MES). A machine needs a preventive maintenance at regular intervals.

In view of product definition, a job has its own process flow in a wafer fab. There are hundreds or thousands of jobs in a wafer fab at the same time. They may belong to different product editions. If the process flow of each product edition is stored in MES, the data storage will be huge. And the process flow of a product edition is difficult to be modified. In addition, a process flow is composed of hundreds of operation sets. An operation sets includes several operations. An operation is corresponding to several pairs of machine and recipe. An operation of one job may be the same with that of other jobs. So we only need storing these operations in MES and combining them into different process flows.

In view of scheduling of a semiconductor wafer fab, its objective is to reasonably utilize the machines and resources in the fab to achieve better operational performance and meet the requirements of the jobs' process flows and other constraints as well. In other word, the scheduling behaviour is responsible for arranging right resources for an operation at right time. There are three main scheduling ways in a semiconductor wafer fab, i.e., static scheduling (usually called sequencing), dynamic dispatching and rescheduling. Static scheduling is to determine the processing order and time for specified operations before the real processing begins. Dynamic scheduling is to determine next processing route of a job or a processing priority of a job on a machine according to current states of a fab in a real-time style. Due to its uncertain manufacturing environment, an existing sequencing plan for a wafer fab may be not applicable to implement. Then the sequencing plan must be modified or regulated to be suitable to the real production environment. This behaviour is called rescheduling.

Based on the analysis results as above, the scheduling of a semiconductor wafer fab can be abstracted as four parts, i.e., WIP-centered configurable definition, machine-concerned physical environment, process information recording production information, and scheduling describing detail algorithms. Then the general structure model of scheduling of a semiconductor wafer fab can be divided into static part and dynamic part (shown in Fig.1).

The static part defines static information of a semiconductor wafer fab, including physical (machine-centred) and configurable (product-centred) definition. The dynamic part defines the information used during the production, including process information (e.g., release plan and sequencing plan) and scheduling related information (e.g., scheduling algorithms). Configuration definition regulates the production on the physical machines by defining the process of WIP. It also offers reference to the scheduling algorithms. Process information provides information (usually a sequencing plan) to machines to facilitate their production processes. On the contrary, the processing on a machine generates process information, too. In addition, process information acts as the input or feedback of scheduling algorithms.

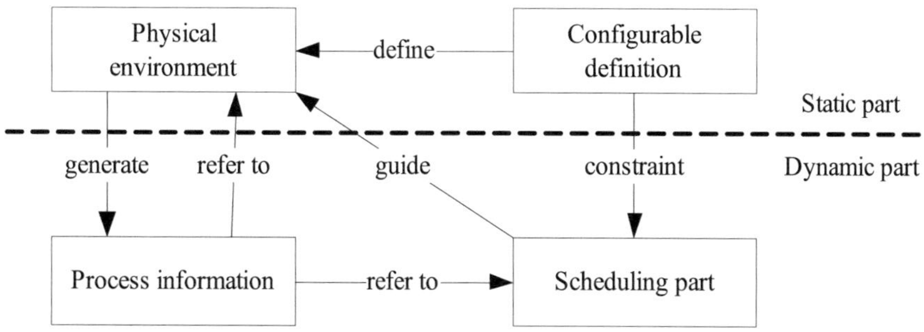

Fig. 1. Main part of General structure model (GSM).

Transferring those parts to its concept layer, a GSM of scheduling of semiconductor wafer fabs is also defined as four layers, i.e., configurable definition layer, physical layer, process information layer and scheduling layer. The components in each layer and the relations between these layers are shown in Fig.2.

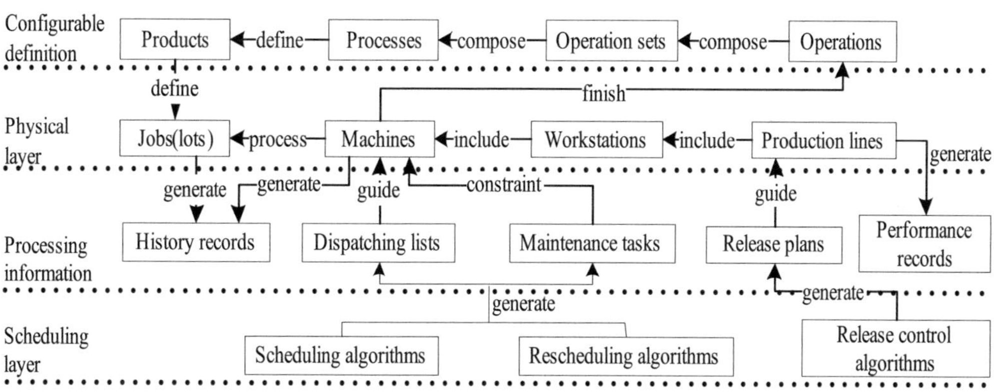

Fig. 2. General structure model (GSM) and its definition at each layer.

The configurable definition layer manages process information of the products of a semiconductor wafer fab, such as products, processes, operation sets and operations. The minimum unit is the operation. Multi-operations compose an operation set, and multi-operation sets constitute a process. Every product has its own process specifications.

The physical layer records resource information of a semiconductor wafer fab, such as its layout and machines, the process abilities and state information of these machines, and state information of WIPs.

The process information layer records the data generated from the simulation processes, including input and output information. Input information contains release plans and machine maintenance plans. Output information includes dispatching lists, machine maintenance tasks, performance issues and the work log of the machines.

The scheduling layer contains some algorithms used by the simulation, including release control algorithms, scheduling algorithms and rescheduling algorithms. Release control algorithms generate the release plan during a special period. Scheduling algorithms, applied to the simulation running, generate the dispatching lists and the maintenance tasks for the machines during a special period. Rescheduling algorithms are used to modify the existing dispatching lists or generate new dispatching lists for a special period.

3. Data-based dynamic simulation modelling method

There are many distinctions between different semiconductor wafer fabs. As a result, it is very difficult to abstract their commonness. The traditional simulation modeling method, focusing on a special semiconductor wafer fab, has strong pertinence but less generalness. Correspondingly, the release control and scheduling algorithms used in a traditional simulation model only fit to a special simulation model, which are difficult to be applied to other simulation models to obtain better performance. In addition, it is not easy to evaluate those algorithms' performance under different production environments.

The less generalness of the traditional simulation models can be solved by using the proposed GSM in section 1. As for the strong pertinence, it is noticed that the differences between the semiconductor wafer fabs are their layouts, machines, processes and scheduling methods that can be transferred to data by some ways. If it is built dynamically by using these data, the simulation model is merely dependent on these data, which overcomes the deficiency of the strong pertinence.

Data types for dynamic modelling

The process of dynamic modelling of the scheduling simulation model of semiconductor wafer fabs is to upload data, handle data, and finally organize data into a simulation model with specified structure. Each data unit belongs to some module in a layer of GSM. So the first job to analyse the require data types for dynamic modelling of a simulation-based scheduling model is to analyse the data requirements of each module in GSM.

Based on the definition on GSM, data types required by the simulation-based scheduling model are summarized into Table 1.

Layers	Modules	Required	Data types
Configurable definition layer	Product, process, operation set and operation	Yes	Product data
Physical layer	Job, machine, work centre and wafer fab	Yes	Physical resource data
Process information layer	History records, sequencing plan and performance records	No	
	Release plan and preventive maintenance	Yes	Production data
Scheduling layer	Static scheduling, dynamic scheduling and on-line optimization (i.e., rescheduling)	Yes	Algorithm data

Table 1. Data types required by dynamic modelling.

- Product data

Product data is corresponding to the configurable definition layer in GSM. Because every wafer fab has its own products' definition, the product data is required without doubt for dynamic simulation modelling. Product data includes information related to products and relations between operation sets and machines. The latter defines the relations between physical layer and configurable layer in GSM.

- Physical resource data

Physical resource data is corresponding to the physical layer in GSM. It describes the physical layout of a semiconductor wafer fab. Because different wafer fabs have their own physical layout, the physical resource data is required without doubt for dynamic simulation modelling, too. Physical resource data is the main body of dynamic simulation model. It decides the numerical layout of a wafer fab in computer software. Traditional modelling methods directly set physical resource data into a special simulation model. If there are some changes (e.g., new machines addition) in the layout, the simulation model must be modified. However, data-based modelling is to build the numerical layout of a wafer fab dynamically according to physical resource data. Therefore, if there are changes in the physical layout of a wafer fab, no extra work is required during modelling process. The dynamic data-based modelling method is more flexible and general than traditional modelling methods.

- Process data

Process data, corresponding to the process information layer in GSM, includes the information required by production processes, e.g., maintenance tasks and release plans. It is the precondition to run the simulation. Process data doesn't include information about the historical records, sequencing plans and performance records that are generated during the simulation running process. So these data is unrequired during model uploading process.

- Algorithm data

Algorithm data, corresponding to the scheduling layer in GSM, records the scheduling algorithms applied to a wafer fab. The detailed scheduling algorithms are realized in the simulation model. Algorithm data records the information related to the algorithms, including algorithm names and parameters.

Once these four kinds of data are self-contained, the simulation model of a semiconductor wafer fab can be loaded dynamically successfully. The flexible superiorities of dynamic data-based simulation model comparing to traditional simulation model are as follows.

Firstly, these four data types are relatively independent and can be easily combined to build simulation models of different semiconductor wafer fabs. So the reusability of these data types is enhanced and the workload of simulation system development is seriously reduced.

Secondly, it is much easier to modify a simulation model. To modify a simulation model needs only the modification of the data required by generating the dynamic simulation model.

Dynamic modelling process

In view of traditional simulation modelling method, its structure (i.e., physical layer) is determined at the beginning of building the simulation model. Data is added to the physical layer. So if there are some changes in the physical layer, the data and model need corresponding modification with heavy workload. This kind of modelling process is called static modelling (shown in Fig.3). The data of static modelling will not be uploaded unless the simulation process starts.

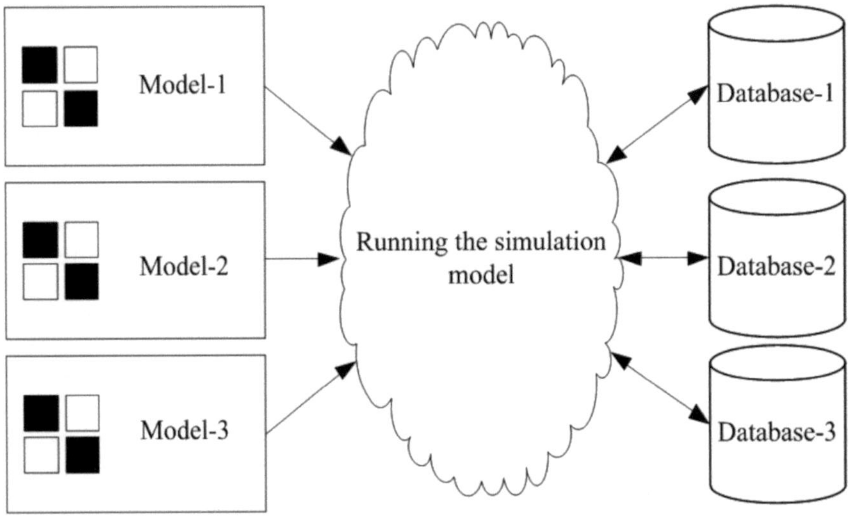

Fig. 3. Static modelling diagram.

In view of dynamic data-based simulation modelling method, data takes an important role during the modelling process. On the contrary, the simulation model is an assistant. All of the changes of the simulation model are realized by changing the data. In other word, the simulation model of the same semiconductor wafer fab may be different due to the changes of its data. This kind of modelling process dependent on the data is called dynamic modelling (shown in Fig.4).

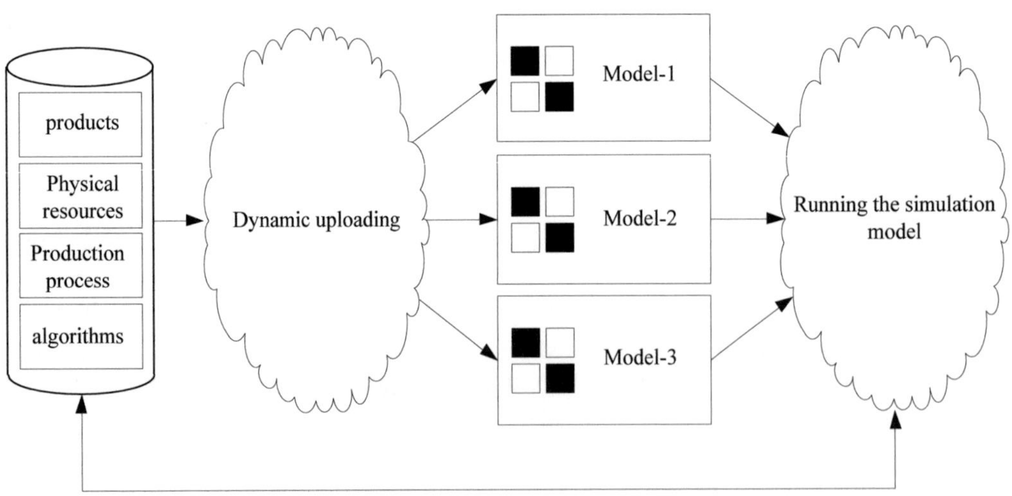

Fig. 4. Dynamic modelling diagram.

The data-based dynamic simulation modeling flow is as follows: load the original data first, then deal with these data to standardize them into structural data required by dynamic modelling, and finally organize these structural data into a simulation model with a specified structure. These data will attribute to some modules of some layers of GSM ultimately.

4. Simulation-based modular scheduling system

Based on the proposed dynamic simulation modeling method in section 3, we design and develop a simulation-based modular scheduling system (SMSS). SMSS integrates six semiconductor wafer fab models, including 4-inch wafer fab (Bl4), 6-inch wafer fab (Bl6), HP24Fab1, HP24Fab2, HP24Fab3 and Mini-FAB. Modular means that SMSS integrates various kinds of scheduling in a wafer fab, such as release control, sequencing, dynamic dispatching and rescheduling methods. Each scheduling method is encapsulated as a module. The modules belonging to the same kind of scheduling can be replaced each other.

The hardware environment of SMSS is Pentium IV 1.5 CPU, 256M memory and 10G disk driver or more. The software environment includes Windows 2000 or XP operation system, .NET Framework 2.0, Microsoft Office 2000 or higher edition, Graphics.Server.NET.v1.1 and simulation platform eM-Plant.

The architecture of SMSS is shown in Fig.5. There are three layers in SMSS, i.e., data layer, software layer and simulation layer.

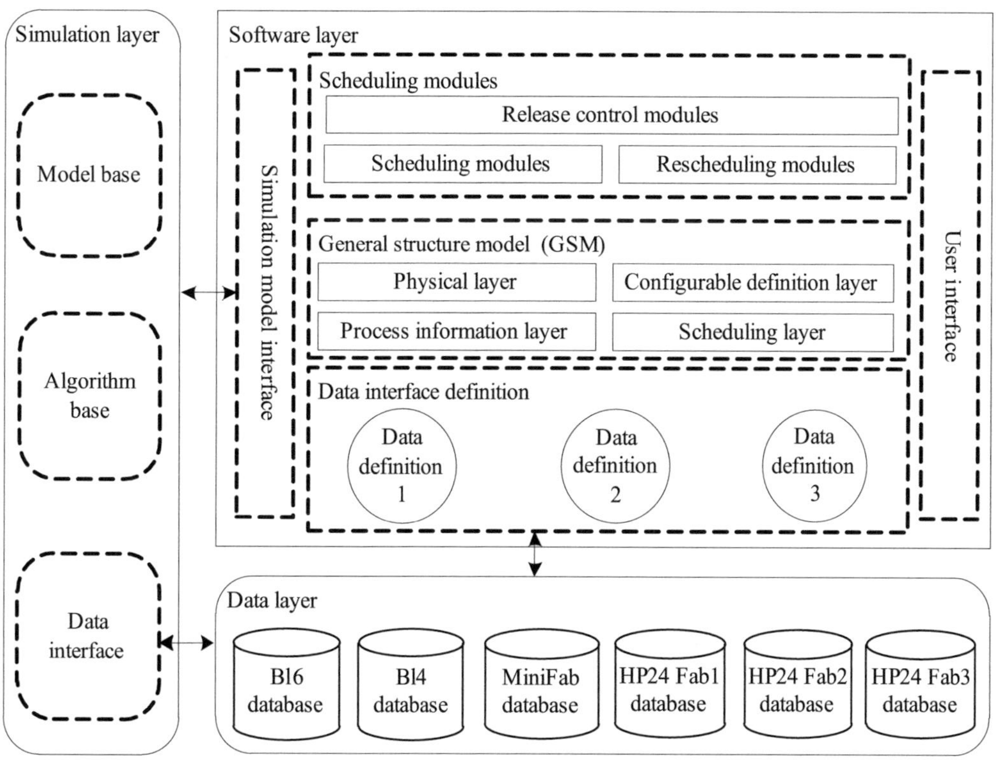

Fig. 5. Overall design of simulation-based modular scheduling system (SMSS).

Data layer

Data layer, realized with an ACCESS database, storages the data of all semiconductor wafer fab models in SMSS. These data are the base to build the basic model in software layer. There are communications between data layer and software layer. There are communications between data layer and simulation layer, too. These communications are enabled through corresponding data interfaces to facilitate free transfer between different semiconductor wafer fab models.

The design principle of data layer is to guarantee identical data structures of six semiconductor wafer fab models. The standard of the identity is the data types in Table 1. To meet the requirements of GSM on data, we build following data-entity tables in the ACCESS database.

- Order table: the attributes include customer, priority, number, due date and etc. The order table is the base of release plans.
- Release table: the attributes include release number, release date, belonging products and etc.
- Maintenance task table: the attributes include the latest start time, the earliest start time, duration time and etc.
- Product table: the attributes include the definition of process flow, the processing time of operation, average cycle time of product, average move step and etc.
- Work-centre and machine table: the attributes include the constitution of work-centre, machine state, machine type, scheduling rule used by machine, processing history record and etc.
- WIP table: the attributes include WIP state, machine queuing, state start time, state finish time and etc.
- Dispatching list table: the attributes include operation start time, operation finish time, job number, machine number and etc.
- Performance table: include all the performance records in a wafer fab.
- Scheduling algorithm table: the attributes include the name of scheduling algorithm and the relations between algorithms and machines.

The relations between above-mentioned data-entity tables and data types in Table 1 are shown in Table 2.

Software layer

Software layer is the core of SMSS. It is used to model a semiconductor wafer fab in a standard way, manage SMSS's modules configuration, and provide a user management interface. Through specified data interface, software layer can read data from data layer. Then it configures the modules in it and sets the parameters of modules. Finally, it can call simulation layer to realize the simulating process of semiconductor wafer fab models.

Data types	Data-entity tables
Product data	Product table
Physical resource data	Work-centre & machine table, WIP table
Production data	Release table, maintenance task table, dispatching list table, performance table
Algorithm data	Scheduling algorithm table

Table 2. Relations between data-entity tables and data types.

Software layer includes five components, i.e., data interface definition, general structure model, scheduling modules, simulation interface and user interface. In addition, we add some functions to make statistics on system data, generate their diagrams and files, and manage the system parameters.

- Data interface definition

Data interface definition is to read data from data layer to software layer and organize these data according to the data types in GSM (shown in Fig.6). Data layer stores data from six semiconductor wafer fab models. Although their data are different, the structures of their databases are the same.

So we can only develop one data interface to read either model. If there are other databases with different structures, we only need to define corresponding data interfaces to integrate these heterogeneous databases easily.

- General structure model

General structure model organizes the data read from data layer into data types described in Table 1 to facilitate scheduling modules to call them quickly.

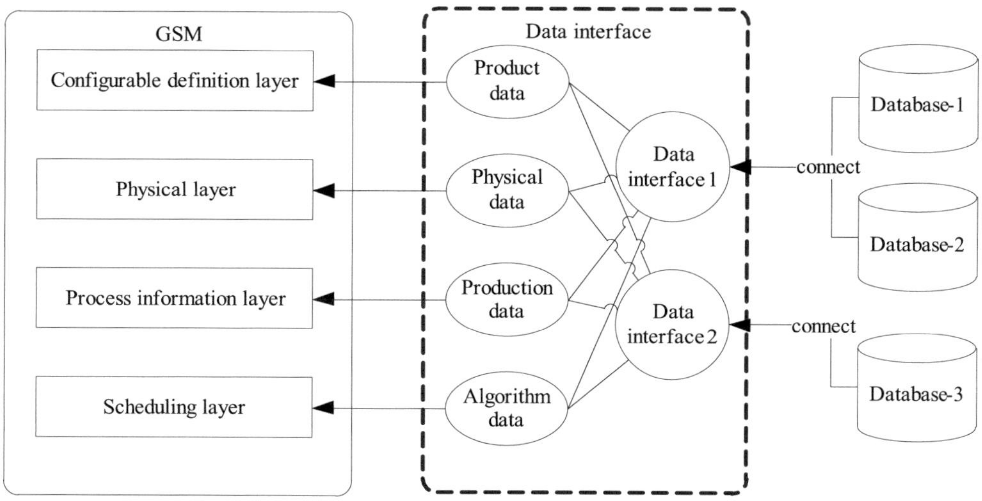

Fig. 6. Data interface diagram.

- Scheduling modules

The scheduling algorithms in MPSS are realized as multiple modules. A scheduling module obtains the algorithm configuration information from user interface, then configures the algorithm's parameter and implements the computation process, finally sends the computation results to the user through the user interface.

- Simulation interface

Simulation interface is the communication interface between software layer and simulation layer. Its functions include: the open, close and save of simulation models, simulation control (i.e., simulation start, pause and reset), and implementation of simulation inner script language.

- User interface

User interface is the interactive interface for user operating software layer. The design principle of use interface is full-open, i.e., the panels for setting parameters are all on the main frame and reducing the number of menus as less as possible.

Simulation layer

The simulation layer is to simulate a current semiconductor wafer fab model. It takes eM-Plant 7.5 as its platform, including a model base, an algorithm base and a data interface.

- Model base

There are six semiconductor wafer fab simulation models in the model base. These models are generated dynamically according to the method introduced in section 3. So the structure, control programs and data interface of these models are identical.

- Algorithm base

There are four parts in algorithm base: model uploading algorithm, simulation model control algorithm, scheduling rules and statistical algorithm. These algorithms are applicable to all simulation models in the model base.

- Data interface

Data interface is used to interact with database to read original data from data layer and write simulation data back to the database.

5. Case study

SMSS can be used as a platform to test the performance of different scheduling algorithms and the proper mix level of scheduling algorithms. Thus, it can be considered as a reference tool for the production decisions.

The workflow of SMSS includes uploading simulation model, generating release plan, generating sequencing plan, dispatching and rescheduling (shown in Fig.7).

Uploading simulation model

The uploading simulation model process is to transfer data from data layer to software layer. Due to huge amount data in semiconductor manufacturing, this process may spend longer time. The uploading time for six simulation models is shown in Table 3.

1. Release control modules

There are four release control modules, i.e., ConNum (the number released to the fab per day is the same), HybridIntelligent (the release plan is generated by an immune algorithm), PredictDueDate (the jobs released to the fab according to the urgent level of their expected due dates) and MultiObjective. Taking Minifab as an example, the performance of each release control modules are shown in Table 4.

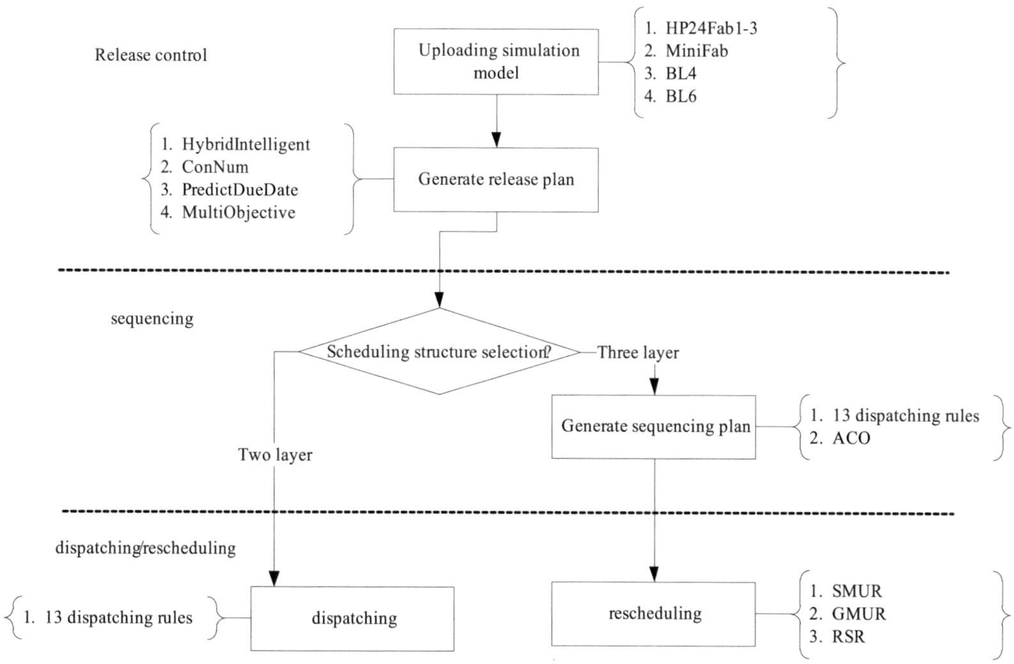

Fig. 7. SMSS workflow and its main modules.

Simulation model	Uploading time (s)
MiniFab	4.1 ~ 4.3
HP24Fab1	23.1 ~ 23.3
HP24Fab1	22.9 ~ 23.2
HP24Fab1	20.3 ~ 20.5
Bl4	110.5 ~ 111.1
Bl6	168.7 ~ 169.6

Table 3. Uploading time of six simulation models.

Release control	Cycle time (min)	Cycle time variance (min)	On-time delivery rate	Throughput (lot/day)	MOV (lot)	WIP (lot)
HybridIntelligent	4780.38	3009.65	0.74	15.17	99	57.17
ConNum	6728.39	3267.84	0.74	15.47	100.22	76.33
PredictDueDate	4902.45	3500.41	0.97	15.27	100.00	59.47
MultiObjective	6580.04	2359.95	0.50	15.67	100.9	72.73
Average	5747.82	3034.46	0.74	15.40	100.03	66.43

Table 4. Performance statistics of release control modules.

2. Scheduling modules

The scheduling algorithms in SMSS include FIFO, EDD, EODD, LPT, SPT, CR, FSVCT, LS, FIFO+, SRPT, SRPT+, SRPT++, FSVL and DBR. The release period is set to 4 months. We select the data in the middle 2 months to implement analysis work. The judge criterion is set to the product of the on-time delivery and the throughput. The former 10 matching of the release control algorithms and the scheduling algorithms are shown in Table 5. Obviously, the matching of HybridIntelligent and DBR is the best decision. It can be seen from the simulation results that SMSS has a strong ability to support impersonal evaluation on different scheduling algorithms and release control modules. It will take an important role on assisting the operation management of complex production environments.

Release control algorithm	Scheduling algorithm	On-time Delivery (%)	Throughput (lot/day)	Judge criterion
HybridIntelligent	DBR	75.86	4.35	3.29991
HybridIntelligent	SPT	77.17	4.23	3.264291
ConNum	DBR	75.88	4.28	3.247664
PredictDueDate	DBR	80.08	4.02	3.219216
HybridIntelligent	FIFO	75.69	4.25	3.216825
HybridIntelligent	SRPT++	75.69	4.25	3.216825
PredictDueDate	EODD	78.05	4.1	3.20005
HybridIntelligent	EODD	75.29	4.25	3.199825
HybridIntelligent	SRPT+	75.29	4.25	3.199825
PredictDueDate	FSVCT	75.59	4.23	3.197457

Table 5. Good matching sequence.

6. Conclusion

This book chapter introduces a simulation-based modular planning and scheduling system of semiconductor manufacturing (SMSS). The main advantages of SMSS are as follows.

1. It adopts a data-based dynamic simulation modelling method to be adaptive to various production environments.
2. It applies a module-based decouple of the algorithms from the simulation models to enhance the simulations' efficiency and make simulation models' to be reused.
3. It offers a better impersonal evaluation platform for further research on the scheduling algorithms.

7. Acknowledgment

This work was supported in part by National Natural Science Foundation of China (No.50905129, No. 61034004) and Program for Scientific Innovation of Shanghai (No.09DZ1120600, 10DZ1120100). The authors would like to thank the editors and anonymous reviewers for their valuable comments.

8. References

Baudouin, M.; Ruberti, C.; Arekion, J. & Kieffer, J. P. (1995). Decision Support System Based on a Factory Wide Information Integrated System and Discrete Event Simulation to

Help Solve Scheduling Problems in a Semiconductor Manufacturing Environment, Proceedings of IEEE Symposium on Emerging Technologies & Factory Automation, pp. 437-445, Paris, France, October 10-13, 1995

Chen, T. & Wang, Y. C. (2009). A Nonlinear Scheduling Rule Incorporating Fuzzy-Neural Remaining Cycle Time Estimator for Scheduling a Semiconductor Manufacturing Factory-A Simulation Study. *International Journal of Advanced Manufacturing Technology*, Vol.45, No.1-2, (November 2009), pp. 110-121, ISSN 0268-3768

Chou, F. D.; Wang, H. M. & Chang, P. C. (2008). A Simulated Annealing Approach with Probability Matrix for Semiconductor Dynamic Scheduling Problem. *Expert Systems with Applications*, Vol.35, No.4, (November 2008), pp. 1889-1898, ISSN 0957-4174

Horn, S.; Weigert, G.; Schönig, P. & Thamm, G. (2006). Application of Simulation-Based Scheduling in a Semiconductor Backend Facility, *Proceedings of ESTC 2006 - 1st Electronics System integration Technology Conference*, pp. 1122-1126, ISBN 978-142-4405-52-7, Dresden, Saxony, Germany, September 5-7, 2006

Jeng, W.D. & Tsai, M.S. (2010). Scheduling Semiconductor Final Testing a DBR Based Simulation Model, *Proceedings of 40th International Conference on Computers and Industrial Engineering: Soft Computing Techniques for Advanced Manufacturing and Service Systems*, ISBN 978-142-4472-95-6, Awaji, Japan, July 25-28, 2010

Kim Y. D., Shim S. O., Choi B. & Hwang H. (2003). Simplification Methods for Accelerating Simulation-Based Real-Time Scheduling in a Semiconductor Wafer Fabrication Facility. *IEEE Transactions on Semiconductor Manufacturing*, Vol.16, No.2, (May 2003), pp.290-298, ISSN 0894-6507

Koyuncu, N.; Lee, S.; Vasudevan, K. K.; Son, Y. J. & Sarfare, P. (2007). DDDAS-Based Multi-Fidelity Simulation for Online Preventive Maintenance Scheduling in Semiconductor Supply Chain, *Proceedings of Winter Simulation Conference*, pp. 1915-1923, ISBN 978-142-4413-06-5, Washington, DC, United states, December 9-12, 2007

Li, W. & Mason, S. J. (2007). Comparison of Simulation-Based Schedule Generation Methodologies for Semiconductor Manufacturing, *Proceedings of IIE Annual Conference and Expo 2007 - Industrial Engineering's Critical Role in a Flat World*, pp. 1387-1392, Nashville, TN, United states, May 19-23, 2007

Liu, A.; Yang, Y.; Liang, X.; Zhu, M. & Yao, H. (2010). Dynamic Reentrant Scheduling Simulation for Assembly and Test Production Line in Semiconductor Industry. *Advanced Materials Research*, Vol.97-101, (2010), pp. 2418-2422, ISSN 1022-6680

Peyrol, E.; Floquet, P.; Pibouleau, L. & Domenech, S. (1993). Scheduling and Simulated Annealing Application to a Semiconductor Circuit Fabrication Plant. *Computers and Chemical Engineering*, Vol.17, No.Suppl, (Oct 1993), pp. 39-44, ISSN 0098-1354

Rajesh P. & Sen A. P. (2004). Simplification Strategies for Simulation Models of Semiconductor Facilities. *Manufacturing Technology Management*, Vol.15, No.7, (July 2004), pp. 618-625, ISSN 1741-038X

Ralph M. ; Christos A. & Leon F. M. (2007). Automatic Generation of Simulation Models for Semiconductor Manufacturing, *Proceedings of 2007 Winter Simulation Conference*, pp. 648-657, ISBN 978-142-4413-06-5, Washington, DC, USA, December 9-12, 2007

Ramírez-Hernández, J. A.; Crabtree, J.; Yao, X.; Fernandez, E.; Fu, M. C.; Janakiram, M.; Marcus, S. I.; O'Connor, M. & Patel, N. (2010). Optimal Preventive Maintenance Scheduling in Semiconductor Manufacturing Systems: Software Tool and Simulation Case Studies. *IEEE Transactions on Semiconductor Manufacturing*, Vol.23, No.3, (August 2010), pp. 477-489, ISSN 0894-6507

Ramírez-Hernández, J.A. & Fernandez, E. (2010). Optimization of Preventive Maintenance Scheduling in Semiconductor Manufacturing Models Using a Simulation-Based

Approximate Dynamic Programming Approach, *Proceedings of the IEEE Conference on Decision and Control*, pp. 3944-3949, ISBN 978-142-4477-45-6, Atlanta, GA, United states, December 15-17, 2010

Shi L. ; Zhang X. & Li, L. (2008). Simulation and Analysis of Scheduling Rules for Semiconductor Manufacturing Line, *Proceedings of the IEEE International Conference on Industrial Technology*, ISBN 978-142-4417-06-3, Chengdu, China, April 21-24, 2008

Sivakumar, A. I. & Gupta, A. K. (2006). Online Multiobjective Pareto Optimal Dynamic Scheduling of Semiconductor Back-End Using Conjunctive Simulated Scheduling. *IEEE Transactions on Electronics Packaging Manufacturing*, Vol.29, No.2, (April 2006), pp. 99-109, ISSN 1521-334X

Tang, C. H.; Qian, Y. L.; Zhu, J. & Yan, S. J. (2010). A Scheduling Method in Semiconductor Manufacturing Lines Based on Genetic Algorithm and Simulated Annealing Algorithm, *Proceedings of ICINA 2010 - 2010 International Conference on Information, Networking and Automation*, pp.1429-1432, ISBN 978-142-4481-05-7, Kunming, China, October 17-19, 2010

Thompson, M. (1996). Simulation-Based Scheduling: Meeting the Semiconductor Wafer Fabrication Challenge. *IIE Solutions*, Vol.28, No.5, (May 1996), pp. 30-34, ISSN 1085-1259

Weigert, G.; Klemmt, A. & Horn, S. (2009). Design and Validation of Heuristic Algorithms for Simulation-Based Scheduling of a Semiconductor Backend Facility. *International Journal of Production Research*, Vol.47, No.8, (January 2009), pp. 2165-2184, ISSN 0020-7543

Wein L M. (1988). Scheduling Semiconductor Wafer Fabrication. *IEEE Transactions on Semiconductor Manufacturing*, Vol.1, No.3, (August 1988), pp. 115-130, ISSN 0894-6507

Werner, S.; Horn, S.; Weigert, G. & Jähnig, T. (2006). Simulation Based Scheduling System in a Semiconductor Backend Facility, *Proceedings of Winter Simulation Conference*, pp. 1741-1748, ISBN 978-142-4405-01-5, Monterey, CA, United states, December 3-6, 2006

Zhang, H.; Jiang, Z.; Guo, C. & Liu, H. (2006). An Extended Object-Oriented Petri Nets Modeling Based Simulation Platform for Real-Time Scheduling of Semiconductor Wafer Fabrication System, *Proceedings of IEEE International Conference on Systems, Man and Cybernetics*, pp. 3411-3416, ISBN 978-142-4401-00-0, Taipei, Taiwan, October 8-11, 2006

Zhang, H.; Jiang, Z. & Guo, C. (2007). Simulation Based Real-Time Scheduling Method for Dispatching and Rework Control of Semiconductor Manufacturing System, *Proceedings of IEEE International Conference on Systems, Man and Cybernetics*, pp. 2901-2905, ISBN 978-142-4409-91-4, Montreal, QC, Canada, October 7-10, 2007

Zhang H. ; Jiang Z. & Guo C. (2009). Simulation-Based Optimization of Dispatching Rules for Semiconductor Wafer Fabrication System Scheduling by the Response Surface Methodology. *International Journal of Advanced Manufacturing Technology*, Vol.41, No.1-2, (March 2009), pp. 110-121, ISSN 0268-3768

Zhang, T. Z. & Wang, Y. P. (2008). Simulation Research on Production Scheduling of Semiconductor Probing System, *Proceedings of 2008 International Conference on Wireless Communications, Networking and Mobile Computing*, ISBN 978-142-4421-08-4, Dalian, China, October 12-14, 2008

8

Production Scheduling on Practical Problems

Marcius Fabius Henriques de Carvalho and
Rosana Beatriz Baptista Haddad
Pontifícia Universidade Católica de Campinas-PUC-Campinas,
Centro de Tecnologia da Informação Renato Archer-CTI,
Brazil

1. Introduction

Management´s desire to be more competitive and to increase profits through manufacturing is evident. Customer responsiveness, increased output, lower manufacturing costs, better quality, short cycle times, bottleneck control and operational predictability, among many other themes, are hot issues on manager´s minds. The management of manufacturing processes is a complex problem that´s objective is to sell goods and services to the market-place, through internal production resources and supplier agreements and capabilities. It is therefore advisable to structure the solution of the problem hierarchically, considering different aggregation levels of information and decisions, Figure 1.

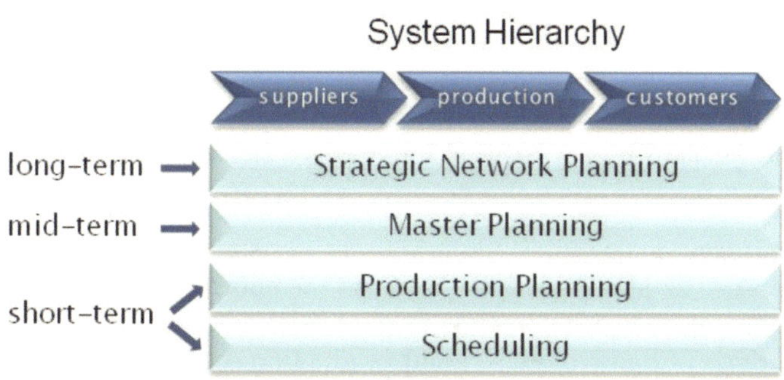

Fig. 1. Production planning management hierarchy

Effective enterprise management decisions at any level must be supported by modern tools and techniques. Both hierarchy and tools can offer tremendous potential advantages when adequately integrated into the search for a good solution. In **long-term operation planning** the strategic business issue, generally looking five or more years ahead, aims to provide for the long-term capacity requirements and resource allocations necessary to meet the future organizational objectives. This is done by planning for capacity changes in line to meet plans for new products/services, technologies and markets (Hill, 1991). Strategic decisions search for two kinds of production flexibility, flexible technology and flexible capacity, to meet long term demand fluctuations. Each strategy is underpinned by a set of operations decisions on technology level, capacity amount, production quantity, and pricing strategy evaluation determines how market uncertainty, production cost structure, operations timing, and investment costing environment affect a firm´s strategic decisions (Yang, Cheng, 2011).

At this level, products with similar production costs and seasonality are grouped together and considered as one family. This procedure subsidizes purchasing, resource development, production and customer service policies and supports the generation of a long-term production plan to guide broad courses of actions that an organization needs to follow in order to achieve its objectives, (Bitran et al 1981,1982; Ozdamar, et al 1998).

The **medium-term** or aggregated planning covers a period of up to two years. The technical feasibility of implementing the strategic level decisions, taking into account time and available resources, are verified. Production plans for product families sharing the same setup costs, that reduce the final production cost, are generated. At this level the data are less aggregated than they are at the strategic level. This makes it possible to generate a plan with a gross level of overtime or subcontract production capacity for the medium-term horizon considered, for example, a few weeks, (Caravilla & Souza, 1995). At this level, the production plan extends beyond the enterprise resources to consider downstream activities as distribution, sales and inventory based on customer market information and upstream activities as supplier contract and constraints, while observing all relevant internal constraints. The production plan connects sources and sinks through material flow equations and is usually driven by economic objectives (Mutairi, 2008).

In the **short term** the production planning data is broken down into tasks to be performed in the short term, for example, a day or a few days. Detailed demand and processes are considered. At this level, sequences are decided on the factory floor, which generally consider several production resources responsible for processing items at the various production stages allocated to each different product (Karimi, B. et al. 2003; Carvalho & Silva Filho 1996). On a fictitious factory floor, dedicated to the production of only one item, with infinite availability of inputs and production resources, it should be relatively simple to coordinate the supply of raw material with the production of components to meet the demand requirement. However, in practice this situation does not occur. Real systems are complex environments where various items compete for limited capacity of the available resources.

Production scheduling deals with operational decisions at each plant, such as the machine production sequence. Its purpose is to transform the short term production plan, utilizing the available resources over a given time horizon into a schedule useful for all operations within a time horizon of a few days. A good production schedule is one that has the ability to synchronize the job chain network, starting on projected and updated completion dates.

The scheduling problems in general can be modeled using continuous or discrete time models. This chapter discusses the discrete time models. They are based on two features. First, the scheduling horizon is divided into a finite number of time intervals with predefined duration. Second, the events are allowed to happen only at the boundaries of the discrete intervals. Hence, the complexity of the problem is reduced and the solution becomes less cumbersome.

Production scheduling methods can be further classified as static scheduling and dynamic scheduling. Static operation scheduling is performed during the compilation of the application. Once an acceptable scheduling solution is found, it is deployed as part of the application image. In dynamic scheduling, a dedicated system component makes scheduling decisions on-the-fly (Wang, Gong, Kastner, 2008).

This chapter is specifically related to the discrete time and static production scheduling problems. Scheduling is typically driven by feasibility and focuses on short-term time horizon. The goal of scheduling is to orchestrate an optimized behavior of a resources-limited system over time, considering the predicable, uncertain and dynamic variables of a real system. Although the research community has considered the scheduling as a solved problem, such a conclusion presumes a rather narrow and specialized interpretation of the problem and ignores much of the process and context in practical environments. The design of more tightly integrated planning and scheduling in real systems is a question that still requires research to reach a practical solution. This theme is considered in this text through the analysis of two auto-part industry applications.

2. Model definition

The real world can be represented by models in order to better understand and propose a solution for some of its particular problems. The models should express the decision maker´s needs and present results in a way that they can be easily read and understood (Voβ and Woodruff, 2006). Also, to find a solution to the problem, it is necessary to consider the level of detail to be included in the model, and the computational effort that must be made to solve it.

For example, by definition, we always prefer a better value of the objective function. However, having the best may not be so important if its cost is high and another solution that satisfies the constraints is enough for the purpose of the decision. In another case, the objective can be of considerable importance and the best solution should be pursued. In either case, we must find a balance between the difficulty in obtaining the result and the desired objective.

In order to define the model, we have to analyse the problems and characterize them according to a variety of aspects and classification criteria. One aspect is the consideration of an integer variable (Wu at. al. 2011) or not since the greater the number of integer variable the harder it will be to solve the problem. But, in some cases it is sensible to require some variables to take on the integer value (Voβ and Woodruff, 2006).

For example, if the product is of a high value, with low production intensity, it does not make sense to talk about a partial product. But if the product is of a low value and higher production intensity, as in the auto part production industry, the integer consideration is not necessary. Also, if the setup time is very high it is necessary to work with integer variables. If however the enterprise is in the middle of a supply chain and receives large weekly orders it may be preferable to process the entire order rather than splitting it into a small one.

Nothing that has anything to do with planning is truly linear, though non-linearity in the variables can make the solutions harder to obtain. But, to integrate the planning model with scheduling a linear model is enough in order to obtain the necessary solution (Voβ and Woodruff, 2006).

Infinite scheduling allows more than one job to be scheduled for production within a limited capacity resource at the same time. When many jobs are scheduled at the same time, which is often the case, the same limited resource is in a state of scheduled chaos. The only alternative is a classic scenario: supervisors and production control personnel must take hands-on control of the shop floor scheduling.

A Finite schedule is one that makes sure that simultaneous schedule contention for the same capacity resource is avoided. Starting from a non-capacitated schedule, a temporal capacitated schedule can be met by rescheduling, following some heuristics, Figure 2.

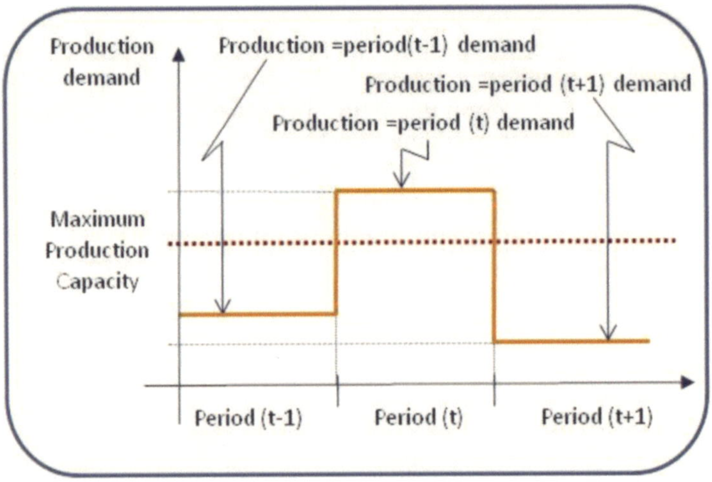

Fig. 2. Reallocation of uncapacitated scheduling.

One way of rescheduling is anticipating or postponing a production order according to priority rules. For example, on a shop floor that produces for both make-to-stock and make-to-order, a "make to stock" job will have lower priority than a "make to fill an order" job. However, a question arises in a system where a temporal solution is needed for multiple customer orders of differing sales values, simultaneously competing for multiple resources. How can an order in this complex environment be prioritized?

Also, distinction can refer to deterministic versus stochastic models. While in deterministic models all data are known in advance, in stochastic models data are based on distributions or on a measure of uncertainty (Voβ and Woodruff, 2006). Static models assume that parameter values do not change over the planning horizon (e.g. a continuous demand at the same rate in every period) while dynamic models allow for variations. The planning horizon can be assumed to be finite or infinite. Costs to be considered are holding costs, setup costs, or production costs.

Production scheduling models can consider one or multiple products. The multi-product consideration implies the development of a multiproduct production plan in a scarce resource environment. The problem can also be single stage or multiple stage and all the above problems can be represented by a capacitated model that recognizes that the capacity of each resource is defined by a finite number or amount. In the situation where there is not enough capacity, certain decisions have to be made, such as those regarding backordering or loss of sale policy.

In any of the above situations, the development of a model must take into account the following points:

- The model must be simple;
- It is advisable to avoid the development of mega-models;
- It is advisable to use similarities and analogies

In the practical application, one can either apply an approximate method that delivers a good solution in an acceptable time or an optimization procedure that yields a globally optimal solution, solving both timing and resource constrained scheduling problems, but requiring a very high computing time (Jain and Meeran, 1999). Another option can even be to integrate both procedures, taking specialized heuristic to treat selected constraints and objectives in an ad hoc manner.

Generally speaking, this would mean taking advantage of a problem-specific engineering solution to obtain a result that meets a given application's requirements. There continues to be great need for research into techniques that operate with more realistic problem assumptions (Smith, 2003). To exemplify the above concepts the following section discusses a number of modeling aspects in more detail.

3. The scheduling problem

Production scheduling can be defined as the allocation of available production resources over time to best satisfy some set of criteria. Scheduling problems arise whenever a common set of resources (labour, material and equipment) must be used to make a variety of different products during the same period of time (Rodammer, White, 1988). The objective of scheduling is to efficiently allocate resources over time to manufacture goods such that the production constraints are satisfied and the production costs are minimized. This involves a set of tasks to be performed, and the criteria may involve both tradeoffs between early and late completion of a task, and between holding inventory for the task and frequent production changeovers (Graves, 1981).

Operations require machines and material resources and must be performed according to a feasible technological sequence. Schedules are influenced by diverse factors such as job priorities, due-date requirements, release dates, cost restrictions, production levels, lot-size restrictions, machine availabilities, machine capabilities, operation precedence, resource requirements, and resource availabilities.

To illustrate a production schedule, consider a one-product, three-machine job-shop scheduling problem, shown in Figure 3. The simplest scheduling occurs when we have unlimited capacity resources for the given application while trying to minimize its latency.

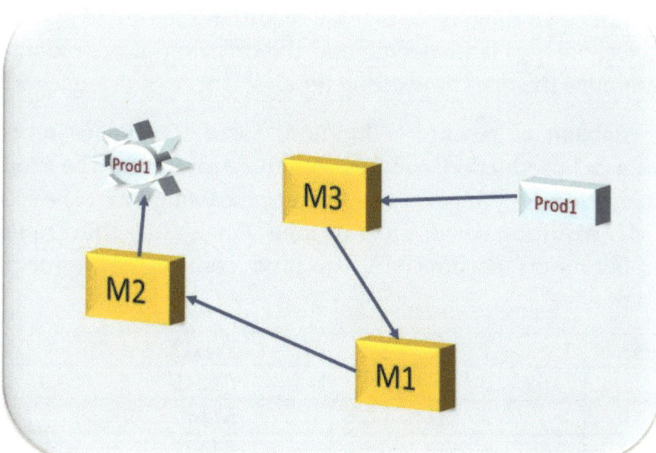

Fig. 3. One-product multi-stage production system

For this problem, simply solve it is to schedule an operation as soon as all of its predecessors tasks have completed, which gives it the name As Soon As Possible. It is closely related with finding the longest path between an operation and the demand due date. The first observation is that the maximum capacity of this system is equal to the lesser of the maximum capacities of the resources.

Now consider Figure 4, with the same three machine system, processing three different products under distinct production sequences. In this case, the maximum productive capacity of the system depends on the mix and sequence of production, the rate of resource sharing and the time profile of demand. Therefore, the production capacity of any multiproduct, multistage system is dependent on the mix of the demand and on the production schedule.

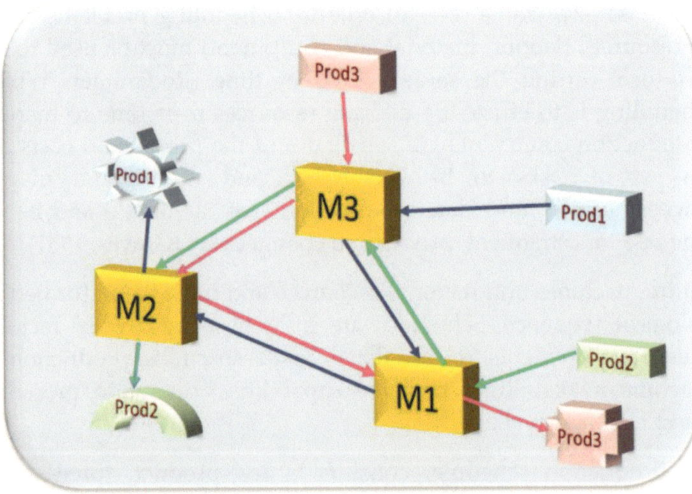

Fig. 4. Multi-stage multi-period multi-product production system

This problem grows in complexity when considering various time periods for manufacturing resource planning. Note that the required ordering of operations within each job (the sequence technology) is preserved and the ordering of operations on the machine was selected to minimize the total production time.

To illustrate the problem of resource allocation, Table 1 provides routes data and the processing time of a cell production, composed of three machines. The Prod1 first operation occurs on the machine M3, and requires one unit of processing time. The second operation is on the machine M1, requiring seven units of time. Finally, the third operation requires 6 units of time and occurs on machine M2. The production sequence for products 2 and 3 follow the same logic.

Operation Sequency	Prod1	Prod2	Prod3
1	1/M3	6/M1	3/M3
2	7/M1	5/M3	6/M2
3	6/M2	3/M2	10/M1

Table 1. Three product production sequence

One of the Gantt charts for production allocation of the three products is shown in Figure 5. It is assumed that all items are available at the beginning of the process and that operations are not shared.

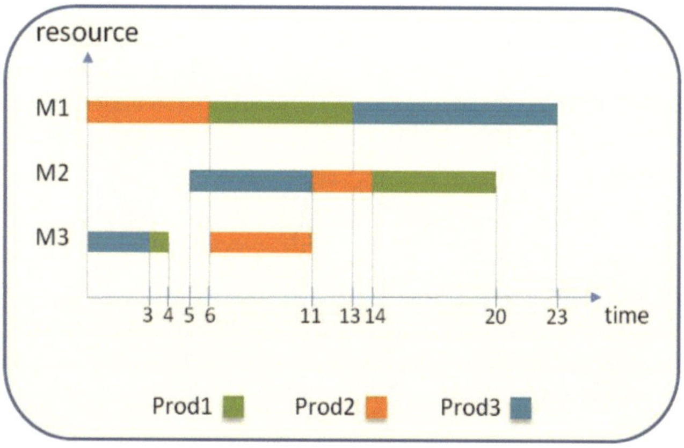

Fig. 5. Gantt chart for three products production allocation

However, when considering a real problem it is not enough to simply have a heuristic to allocate the production of the items to their machines as shown above. It is essential to optimize the sequencing of production in order to properly harness the available resources, avoiding as much as possible the use of overtime, hiring services and loss of orders due to lack of unavailable capacity.

4. Starting the production planning and scheduling with a MRP model

Enterprise Resource Planning (ERP) is one of the industrial practices employed to support financial decisions, quality control, sales forecast and manufacturing resource planning, among other fields (Gershiwin, 1986; Clark, 2003). ERP proposes an integrated solution for the whole enterprise with modules that cover areas such as: Production, Accounts, Finance, Commercial, Human Resources, Engineering and Project Management.

One ERP module is the Manufacturing Resource Planning (MRP), which supports production planning and scheduling decisions. It establishes the quantities and due dates for the items to be manufactured or assembled as well as determining the production resource needs.

The MRP model comes from a production planning perspective rather than an optimization perspective. This is a very practical model and can be used as a basis for even more sophisticated models. An interesting point in MRP is that it generates integer production quantities, provided that the demands and minimum lot sizes are integers. Computerized planning systems based on MRP have been in use for decades and its logic remains at the heart of the production planning module of many modern Enterprise Resource Planning (ERP) systems.

MRP generates the production orders to be implemented on the shop floor, working as if the shop floor had infinite capacity. It also generates production orders that sometimes overload critical resources. When overload occurs, the manager has to decide, based on his experience, which orders should be placed first, postponed and those that will not be implemented. Therefore, the MRP plan does not guarantee enough capacity to actually carry out its implementation. This plan can also be so unrealistic that it will not be useful (Voβ and Woodruff, 2006).

The MRP works with large lot sizes. The primary reason for this is to ensure that not too much productive capacity is used to changeover from one stock keeping unit (sku) to another. But even with infinite capacity and lot size nervousness the MRP is very useful in industries with changing demand patterns where standard orders can not be used. In addition, MRP models can provide a good starting point for planning and for ordering material (Voβ and Woodruff, 2006). Although MRP is presented as a planning tool, it is also often used as a scheduling tool.

To illustrate the MRP main steps, consider product´s P structure shown in Figure 6. P has a Bill Of Materials (BOM) with three row materials (RM1, RM2, RM3), two manufactured items (M1, M2) and two assemblies (A1, A2).

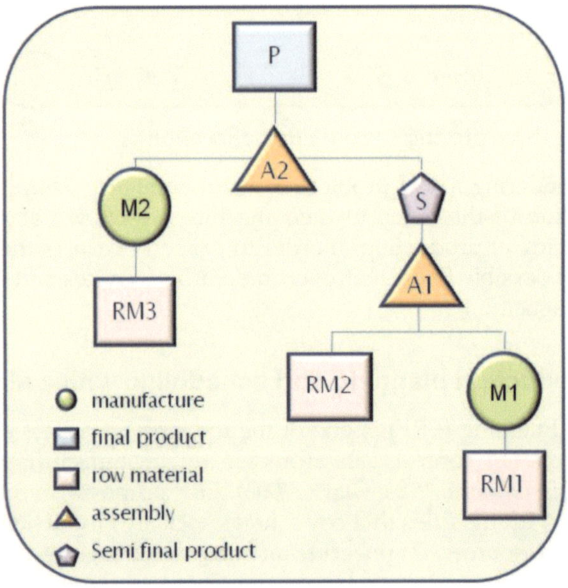

Fig. 6. Structure of a Product

Each one of the components of this BOM has particular characteristics and restrictions, for instance, lead times for acquisition, lot size and initial inventory. The technological restrictions for all items belonging to the Product P BOM are summarized in Table 2.

Item	Lead Time	Minimum Lot Size or Order Quantity	Components	Initial Inventory
RM1	2 days	Order quantity = 200	-	150
RM2	4 days	Order quantity = 450	-	450
RM3	2 days	Order quantity = 150	-	130
M1	1 day	Lot Size = 100	2 RM1	40
M2	2 days	Lot Size = 100	2 RM3	120
S	1 day	Lot Size = 80	3 RM2; 1 M1	70
P	1 day	Lot Size = 100	1 S; 1 M2	90

Table 2. Technological restrictions for the components belonging to the BOM of Product P

	Component P	Day 0	Day 1	Day 2	Day 3	Day 4	Day 5	Day 6	Day 7
	Demand		20	50	40	30	40	50	30
a)	Start of Planned Orders			100			100		
	End of Planed Orders				100			100	
	Inventory Plan	90	70	20	80	50	10	70	40

	Component S	Day 0	Day 1	Day 2	Day 3	Day 4	Day 5	Day 6	Day 7
	Demand		20	50	40	30	40	50	30
b)	Start of Planned Orders			80		80		80	
	End of Planed Orders				80		80		80
	Inventory Plan	70	50	0	40	10	50	0	50

	Component M2	Day 0	Day 1	Day 2	Day 3	Day 4	Day 5	Day 6	Day 7
	Demand		20	50	40	30	40	50	30
c)	Start of Planned Orders			100		100			
	End of Planed Orders					100		100	
	Inventory Plan	120	100	50	10	80	40	90	60

	Component M1	Day 0	Day 1	Day 2	Day 3	Day 4	Day 5	Day 6	Day 7
	Demand		20	50	40	30	40	50	30
d)	Start of Planned Orders		100			100		100	
	End of Planed Orders			100			100		100
	Inventory Plan	40	20	70	30	0	60	10	80

	Component RM3	Day 0	Day 1	Day 2	Day 3	Day 4	Day 5	Day 6	Day 7
	Demand		40	100	80	60	80	100	60
e)	Start of Planned Orders		150		150	150			
	End of Planed Orders				150		150	150	
	Inventory Plan	130	110	10	80	20	90	140	80

	Component RM2	Day 0	Day 1	Day 2	Day 3	Day 4	Day 5	Day 6	Day 7
	Demand		60	150	120	90	120	150	90
f)	Start of Planned Orders		450						
	End of Planed Orders						450		
	Inventory Plan	450	390	240	120	30	360	210	120

	Component RM1	Day 0	Day 1	Day 2	Day 3	Day 4	Day 5	Day 6	Day 7
	Demand		40	100	80	60	80	100	60
g)	Start of Planned Orders		200		200				
	End of Planed Orders				200		200		
	Inventory Plan	150	110	10	130	70	190	90	30

Table 3. Needs for BOM of P components

By defining the product P demand for the next 7 days as 20, 50, 40, 30, 40, 50, 30, it is possible to calculate the demands for all the items in Table 2. Knowing that the MRP is conceived to produce as late as possible, and assuming these components are used only in product P, Product P´s component needs can be determined as in Table 3.

For component P in table 3 a), the initial inventory is 90 (as indicated by Day 0). Because of the demand of 20 units, on Day 1 the inventory plan will be 70 units (90 - 20 = 70). On Day 2 the demand will be 50 units, which will result in 20 units for the inventory plan. On Day 3 there are 20 units missing from the inventory, since the demand is 40 units and the inventory plan shows 20 units. Because of the lead time of 1 day, on Day 2 it is necessary to place an order for P1 with the amount of this order being at least 100 units, which is the minimum lot size of P.

Components RM3 and RM1 (e and g, in Table 3) have double the demand of P, because their incidence is 2 units in components M2 and M1 respectively. The incidence of Component RM2 is three to one product P. The calculations for all components are similar to those made for component P. This is a result from MRP calculations.

The amount of final products in a factory, the technological restrictions of each one of the items in their BOM, and the fact that several of these items belong to more than one BOM are issues that are considered. Thus, the MRP does an efficient job regarding the organization of dates and amounts to be manufactured, assembled or purchased of each one of the necessary items presented in all BOM.

However, this is not enough to avoid the chaos on a shop floor, since MRP does not consider the production capacity of the available resources. It assumes that each one of the resources has an infinite capacity, therefore it is always possible to start the production of an item by considering lead times only.

Besides the fact that resources have limited capacity, equipment is frequently in maintenance, or broken. These aspects must be considered by the manager at each stage of planning. It is the manager´s task to decide which order will be late, which will be outsourced and which will not be carried out.

Considering that some periods are overloaded it seems natural to move the exceed production to those periods that are less busy. But this is not a simple task, with it being necessary to use mathematical tools in order to solve this problem.

5. Mathematical model

The manufacturing process is a pre-defined technological sequence of production activities in the production network. On their way through the production system, the items, raw material, semi-manufactured products and finished products wait in queues for release conditions. They are subjected to fabrication or assembly or transportation operations until they reach the final customer. Figure 7a) represents the pre-defined sequence of production activities for one product and Figure 7b) expands it through three periods of time.

Decisions regarding the quantity of raw material available, production level for each group of machines and demand to be supplied must be made in each time period. Cost and capacity are associated with each production stage (processing, transportation, assembling and storage activities). Costs and capacities can be different for each time period , e.g. at period t, stage S, a machine M has a capacity of 3 units; while at period (t+1) the same machine M at the same stage S can have a capacity of 2 units due to a determined preventive maintenance program.

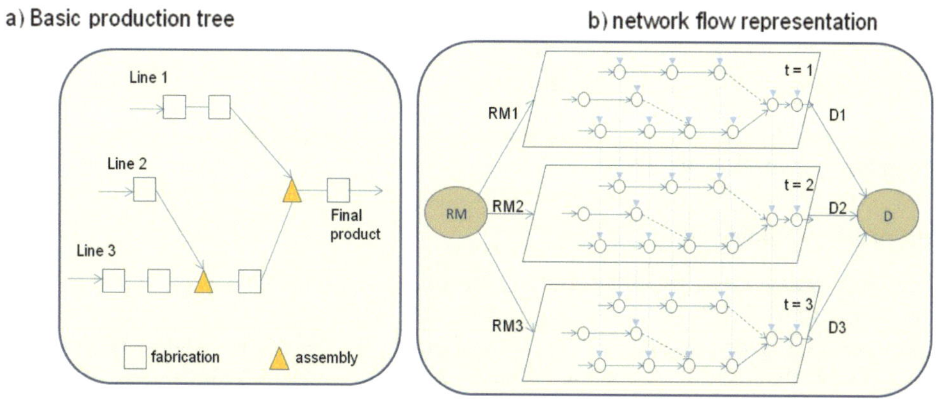

Fig. 7. Sequence of production activities on a production system

5.1 Mono-product system

Bowman (1956) was the first to suggest to solve the production scheduling problem by a transportation method framework, and further, that many transportation problem could be extended to include a multi-time period. Denoting production resources (storage and machines) by arcs and the decision points by nodes, the production routes can be represented as in Figure 7.b). RM represents the raw material storage node and the diverging arcs (RM1, RM2 and RM3) drive the raw material to each period of time. Node D represents the total demand. It results from the sum of flows through arcs D1, D2 and D3.

The flows through the arcs represent the decisions to be made. For example, figure 8 presents the balance equation for a generic node (i,t). The flow x(i-1,t), leaving node (i-1,t) and arriving at node (i,t), through arc (i-1,t) represents the amount to be processed at the production stage (i,t). Flow y(i,t-1) that leaves node (i,t-1) and goes to node (i, t) through arc (i,t-1), represents the amount of material to be stored. The decisions in the node (i, t) is the amount to be produced and the amount to stored.

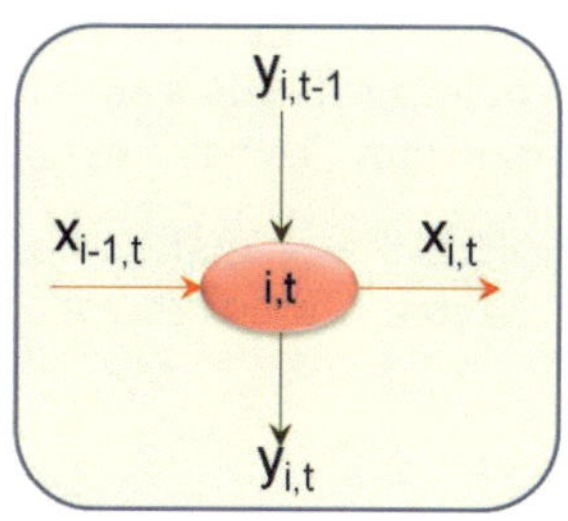

Fig. 8. Balance equation for a generic node

For each production node the balance equation can be written as follows:

$$x_{i-1,t} + y_{i,t-1} = x_{i,t} + y_{i,t}$$

Associating cost and capacity to each arc, the optimisation problem and supplying demand at minimal production cost, can be stated as a network flow model (Carvalho et al., 1999), as shown in Problem P1.

$$\begin{aligned} &\text{Min } f(x) \\ &\text{s.t. } Ax = b \qquad \text{(P1)} \\ &l < x < u \end{aligned}$$

where l and u are the minimum and the maximum capacity associated to a production or storage resource, respectively.

5.2 Multi-product system

In real production systems, several products are processed in the same facility. To handle variations, operation facilities are normally designed with sufficient flexibility to process more than one family of product. The facility capacity must be co-ordinated to guarantee that the sum of the production plan of each individual product does not exceed the production capacity of each facility.

Problem P1 therefore must be expanded to Problem P2, to represent the multi-product production system as shown in (P1).

$$Min \sum_{i=1}^{n} c_i^t x_i$$

$$S.A.: \begin{bmatrix} A_1 & 0 & \dots & 0 & 0 \\ 0 & A_2 & \dots & 0 & 0 \\ \vdots & \vdots & \ddots & \vdots & \vdots \\ 0 & 0 & \dots & A_n & 0 \\ S_1 & S_2 & \dots & S_n & I \end{bmatrix} \begin{bmatrix} x_1 \\ x_2 \\ \vdots \\ x_n \end{bmatrix} = \begin{bmatrix} b_1 \\ b_2 \\ \vdots \\ b_n \\ d \end{bmatrix} \qquad \text{(P1)}$$

$$l \leq x \leq u$$

Where Ai is the incidence matrix of product i, x is the decision variable representing the material flow in the production resource, limited to the lower bound "l" and upper bound "u". Demand is shown by b and raw material availability vector, Si describes the mutual capacity and mutual inventory constraints, called side constraints. The production capacity vector is represented by d. This problem exhibits a special structure that is exploited in the solution considered by PRONET algorithm (Yamakami et al, 2000).

If the production lines are decoupled, the Si matrices are equal to zero and the optimal manufacturing scheduling is reduced to a simple network flow problem.

5.3 Solution algorithm

The large dimension and the very particular structure of Linear Program models associated with the production planning problem motivates the development of algorithms that explore the special structure of the problem. One approach is the Netside Algorithm (Kennington, 1980), but its efficiency is limited to the size of the problems (Carvalho et al., 1999).

Interior Point Methods (IPM) have grown in importance since the positive results obtained by Adler et al. (1989). Nowadays, the IPM for Linear Programming is well established for practical applications and good algorithms are available (Gondzio, 1996; Wright, 1996).

In addition to using the IPM for large-scale problems, faster solutions can be achieved by applying practical knowledge and special characteristics from the real problem throughout their transformation in a mathematical model. The first one is bottleneck management defended by the Theory of Constraints (TOC) (Goldratt and Cox, 1986). According to this theory, a productive system can be divided into two kinds of resources: the bottlenecks and others. Bottlenecks are those resources with limited production capacity, which therefore need special treatment. The planning and scheduling of these resources must be managed carefully in order to meet the demand requirement dates. The decisions regarding other resources are submitted to bottleneck decisions.

The second assumption is the use of mathematical transformation over the constraint matrix. According to Zahorik et al. (1984), "the immense size of these problems and the imprecise nature of many of their costs and demands further suggest that good heuristics may be as desirable as (presumably) more costly optimisation algorithms". As a consequence of this thinking, another way to solve (P1) is to combine both techniques - optimization and heuristic - to attain the best features of each one, as suggested in the followings sections.

5.4 Integration of MRP and linear programming

MRP is an industrial practice largely employed in manufacturing production planning. It is however not enough to completely solve the complex production-scheduling problem, mainly because the basic form of an MRP considers infinite capacity. As a consequence, it does not adequately coordinate the production capacity and raw material time availability with demand requirements and requires additional tools based on expert knowledge of the problem and optimization procedure (Józefowska and Zimniak, 2008)

Temporal co-ordination can be reached by integrating a Linear Programming algorithm with the MRP module of an ERP. MPS generates long term planning for product types as shown in Figure 9.a). Starting from MPS targets, the P1 optimization program assesses the ERP Data Base considering temporal production availability, raw material constraints and product priorities, and generates feasible temporal production scheduling. The temporal co-ordination suggests, when necessary, the anticipation or postponing of demand attainment, according to a pre-defined criterion.

To overcome the dimensionality problem of Linear Programming the model only considers the important production resources and product families. It generates a temporal production scheduling for this simplified system, and due to this, its results need to be refined. MRP, starting from the targets established by the linear model, splits the families into items and explicitly considers the complete production tree and set-up time of each resource. Therefore, Linear Programming and MRP are complementary planning tools. This is because the first one has the temporal visibility of a production problem and the second one has the visibility of the items and the set-up time, and the integration of both allow for an accurate representation of the problem.

An alternative scheme is shown in Figure 9.b) that suggests running the MRP module, and temporally verifying production capacity or raw material constraint violation. When violation occurs, the optimization algorithm is executed to generate a feasible solution. In both approaches, the objective is to maximise the customer service level by co-ordinating the temporal distribution of the production according to the production availability of machines, storage and raw materials.

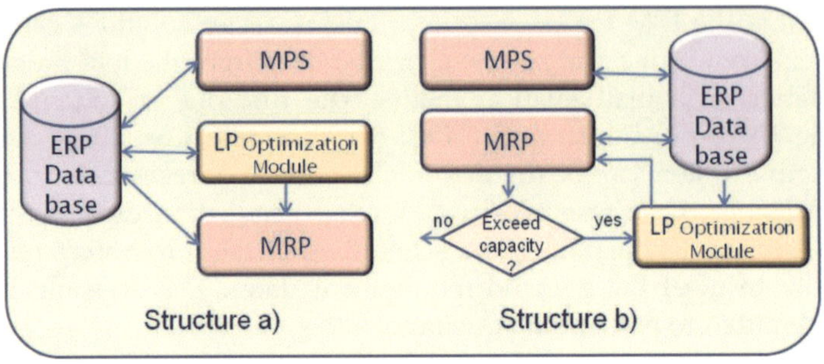

Fig. 9. Alternative structure for Linear Programming and MRP integration.

6. Case study

Solving real problems is not limited to the choice of which tool must be used. Each actual manufacturing system has its own characteristics that will differ from others. It is therefore regarded as unique and requires a detailed analysis of its process in order to define an adequate model. Furthermore, modeling a real situation, whether it is a factory floor, a network of product distribution or a power station, requires abstraction and approximations that transform the actual system into a well-structured mathematical problem.

At this point, the use of empirical knowledge that routinely deals with the actual situation is valuable. This knowledge can be transformed into rules that result in the computational model being as close as possible to the real situation that will be studied. The main steps transform a real problem into a computational model and can be established as:

- To understand the problem to be modelled
- To use empirical knowledge from those who works with the problem
- To transform this knowledge into well-structured rules accepted by the tool and that reflect, as much as possible, the actual situation
- To choose the tool to be used.

6.1 Stamped components industry

The above approach was applied to an Auto-parts enterprise with a commercial ERP installed. This enterprise produces wheels, chassis and stamped components for the main commercial vehicle manufacturers in Brazil, as well as exporting these products to several countries. The objective of the study was to implement a capacitated production schedule in the chassis manufacture. It has a complex production system involving several operations.

This study identified three presses (PR00, PR10 and PR20) as the bottleneck operation in chassis production. These presses with capacities of 3kt, 5kt and 3kt respectively work in parallel to one another. Each item has a predefined fabrication route (Figure 10) with one or two operations in the presses, called operations RF and EF. The 5kt Press can process any item, but items I1, I3 and I4 can be processed only in this particular Press. Since the processing time of each item is around 0.01 hour, while the set-up time for an operation varies from 0.01h to 2.0h, it is interesting to process the largest possible number of items when set-up time is performed. Therefore, a request for an item in a week should be processed as a complete lot.

Practical operations have shown that the processing sequence of items can influence the system set-up time. A heuristic method that aggregates items into families was implemented to take advantage of this particularity. The heuristic method is based on the fact that the set up time of the press operation includes the removal, arrangement or addition of blocks required to process each item in the press. One of these operations will result in a new product.

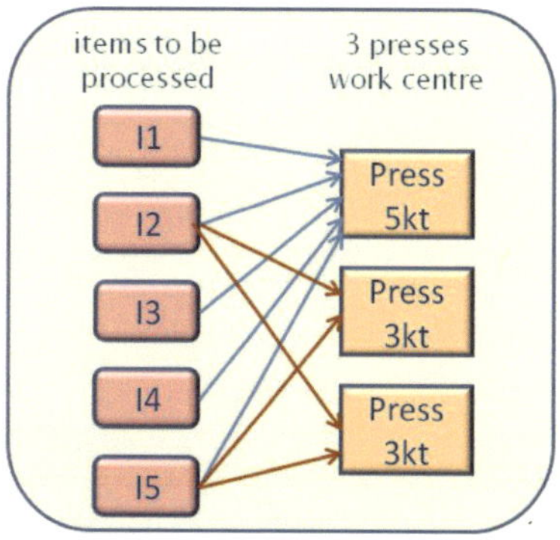

Fig. 10. Fabrication Routes in the Presses

Figure 11 shows the steps for the process preparation sequence within a family, considering that the processing sequence is determined by the number of blocks needed to process each item in the family. The family item with the largest number of blocks starts the process. A new set-up is then performed and the item with the second largest number of blocks of this family is processed.

The heuristic rules to schedule items I141, I142 and I151 are: start with the item in the family that needs the largest number of blocks. After this item is processed, the next in the sequence is the one that needs the second largest number of blocks.

The heuristic method for each family is:

Step 1. Identify the item with the largest number of blocks;
Step 2. process the lot of this item;
Step 3. remove, arrange or replace the necessary blocks to process the next item with the largest number of blocks. Repeat steps 1 to 3 until the last item of the family has been processed;
Step 4. select a new family and go to step 1.

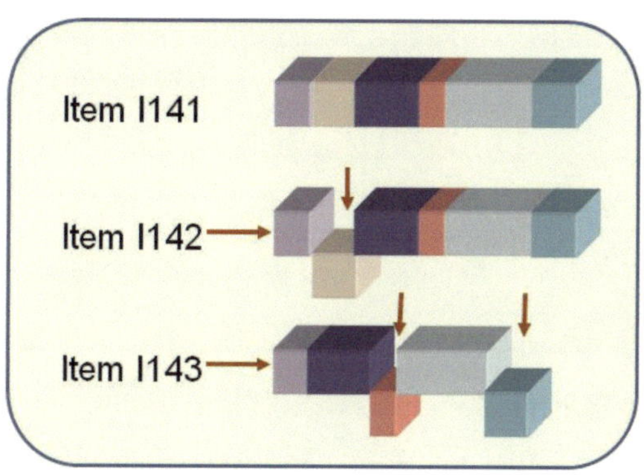

Fig. 11. Items processing sequence

When the processing sequence follows the heuristic method, the set-up time for items of the same family varies from 1 minute to 1 hour and 30 minutes, while set-up time between families is 2 hours. Figure 12 shows a comparison between set-up times for the same group of items. In the first case, items are processed following a FIFO (first in first out) sequence. In the second case the heuristic method grouped the items into families.

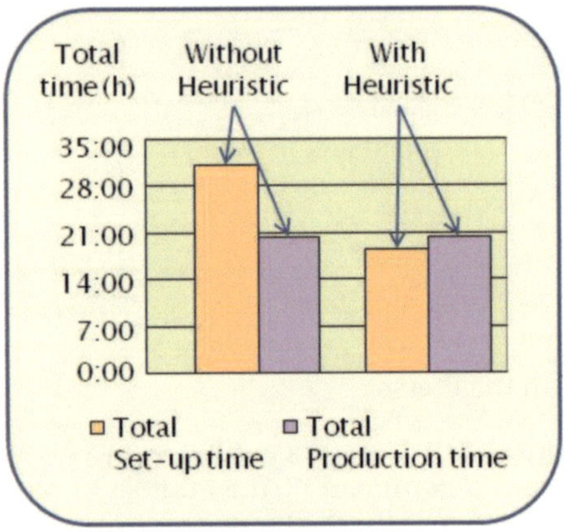

Fig. 12. Set-up times with and without heuristic method

Figure 13 shows the relation between set up time and process time for aggregation of items into families. In this figure items 1 to 4 represent the first family. The second family is made up of items 5 to 8. The set-up time for the first item processed into a family is 2 hours. It is therefore possible to observe that 1, 5, 9, 11, 13, 15, 16, 17, 18, 19, 21 and 23 are the first items in their families, requiring 2 hours for set-up. Families 15, 16, 17 and 18 are composed of just one item.

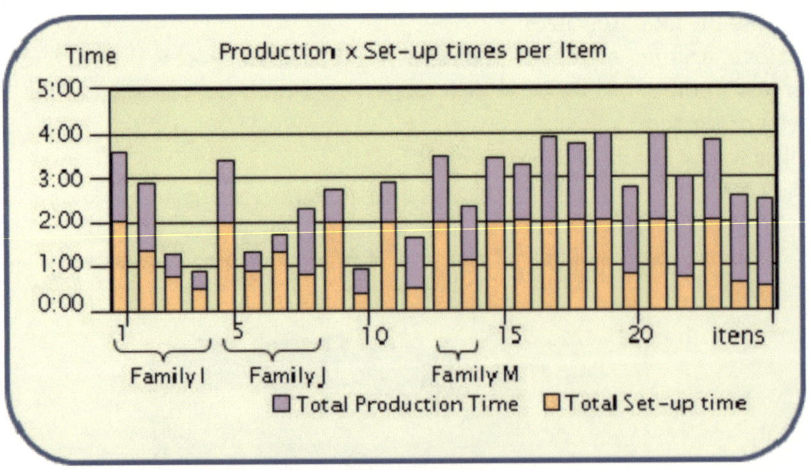

Fig. 13. Families processing sequence

Tables 4, 5 and 6 present actual auto-parts industry data. The first column in Table 4 identifies the families; the second column: items. The third column presents the performed operation - called RF or EF; the fourth column presents the best process sequence for items within a family and column five specifies possible equipment to be used. The suggested sequence in column four was obtained by following the heuristic method.

This table presents two families and their operations in the presses. Four items from family F1 are submitted to RF and EF process operations are (I161, I171, I121 and I131). Three are submitted only to EF process operations (I141, I181 and I191). Four items from family F2 are submitted to RF and EF process operation (I241, I281, I272 and I242) and two are submitted only to EF process operations (I282 and I251).

In daily operations, demand for the complete family of products is unlikely to occur. However, even when there is demand for part of the family the pre-established processing sequence must be observed to minimise the processing time. For example, suppose that in a week the factory had a demand of I272, I241, I171, I121 and I242. The ideal processing sequence should be as shown in Table 4, which means, for Family F1, the item I171 is followed by item I121. For family F2, the sequence should be: I241, I242 and I272 for RF operation and I241, I272 and I242 for EF operation.

	Item	operation	Sequence	Press
Family F1	I161	RF	1	PR00/PR20
	I171	RF	2	PR00/PR20
	I121	RF	3	PR00/PR20
	I131	RF	4	PR00/PR20
	I161	EF	1	PR00/PR20
	I171	EF	2	PR00/PR20
	I131	EF	3	PR00/PR20
	I141	EF	4	PR00/PR20
	I121	EF	5	PR00/PR20
	I181	EF	6	PR00/PR20
	I191	EF	7	PR00/PR20
Family F2	I241	RF	1	PR00/PR20
	I281	RF	2	PR00/PR20
	I282	RF	3	PR00/PR20
	I242	RF	4	PR00/PR20
	I272	RF	5	PR00/PR20
	I251	RF	6	PR00/PR20
	I241	EF	1	PR00/PR20
	I281	EF	2	PR00/PR20
	I272	EF	3	PR00/PR20
	I242	EF	4	PR00/PR20

Table 4. Best process sequence inside a family

The first column of Table 5 is dedicated to families, the second to operation. The fifth column presents the set-up time for exchange items from column three to items in the fourth column. Suppose item I242 is leaving the press in the first operation. Considering the ideal sequence presented in Table 4 for the RF operation, the next item to be processed is I272 with a set-up time equal to 10 minutes -Table 5. For the EF operation, the ideal sequence will be I272 followed by I242 and the set-up time will be 1 hour and 30 minutes.

	Operation	from	to	Set-up (h:min)
Family F1	RF	I171	I121	00:30
	EF	I171	I121	00:40
	RF	I241	I242	01:30
Family F2	RF	I242	I272	00:10
	EF	I241	I272	00:40
		I272	I242	01:30
	:	:	:	:

Table 5. Best process sequence within a family

Table 6 exemplifies the demand for items in 2-week periods.

Item	Period_1	Period_2
I279	-	-
I281	-	-
I171	-	304
I121	-	322
I272	120	-
I241	120	-
I242	26	-

Table 6. Demand per period

Two algorithms have to be developed to transform the data from ERP format to optimizer format: "Families" and "Network", Figure 14. The "Families" module works on the data provided by Table 4 to Table 6 and on the ERP data base (like sequences and bill of material) to create "equivalent families" - EQ, which are the subsets of original families.

The demand size is considered equal to a unit for each EQ. Its processing time is the sum of the processing time of each item with demand in the period plus set-up time (ERP data base and tables 2 and 3).

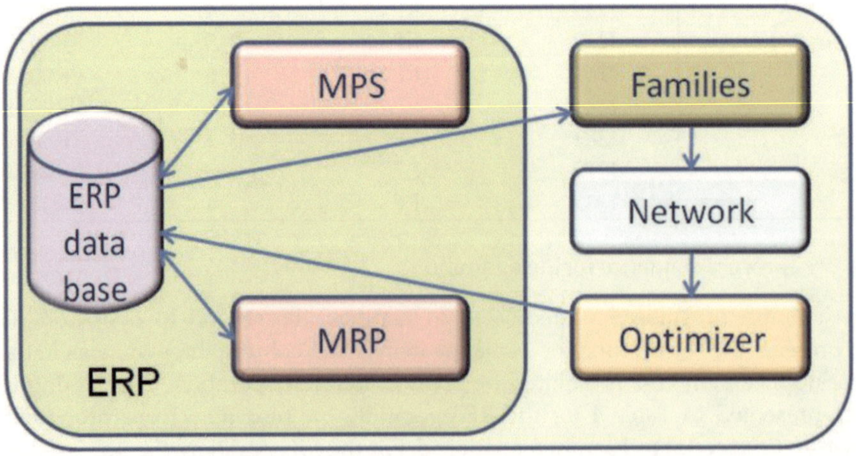

Fig. 14. New configuration of the capacitated manufacturing planning

Table 7 illustrates the procedure for the creation of EQ. In the first column, the families are presented, the second column contains the operations to be processed (RF or EF), and the third column presents the items with demand during the period. The forth column contains set-up times for each item in the family. The first item of each family always needs 2 hours set-up time, in the first or second operations. The set-up time between the first and second item in family F1 at RF operation is 30 minutes and at the EF operation is 40 minutes. The fifth column shows the sum of set-up times of the RF and EF operations for each family. Column 6 shows the demand during the period. The processing time is shown in column seven (50 minutes is necessary to process 70 items of I171). Column eight contains the sum of processing time for the whole family in operations RF and EF. Column nine shows the Equivalent Family name, and column ten the time needed to process the family. E.g.: F1A processing time is (5 hours and 10 minutes + 3 hours and 30 minutes = 8 hours and 40 minutes).

Fam	Op.	Item	Set-up (h:min)	Total Set-up (h:min)	Dem (item)	Proc. time (h:min)	Total Proc. Time (h:min)	EQ	Proc.+ Set-up (h:min)
F1	RF	I171	2:00	5:10	70	0:50	3:30	F1A	5:10+3:30
	RF	I121	0:30		85	1:00			=8:40
	EF	I171	2:00		70	0:45			
	EF	I121	0:40		85	0:55			
F2	RF	I241	2:00	7:50	50	0:30	3:50	F2A	7:50+3:50
	RF	I242	1:30		80	0:40			=11:40
	RF	I272	0:10		80	0:45			
	EF	I241	2:00		50	0:30			
	EF	I272	0:40		80	0:45			
	EF	I242	1:30		80	0:40			

Table 7. Equivalent family creation procedure

Since EQ is seen by PRONET as one item, this guarantees that all items in the family are processed together, thus respecting the best heuristic sequence. This approach enables the use of Linear instead of Integer Programming, overcoming the dimensionality and time consumption problems.

6.1.1 Example

The past production data of the company identifies around 900 items that are grouped into 92 families, organised as in Tables 4 and 5.

Table 8 considers an extract scenario of two periods. Each family and processing time are similar to those presented in table 4. Column 1 shows the Equivalent Family name. Column 2 shows the original MPS schedule for 2 periods. The first 6 lines are related to the first period, and the 11 remaining to the second period. "Press" columns refer to the equipment used to process the family. In this way, family F20A must be processed in press PR20 in the first and second periods. The choice of equipment respects the manager´s priorities.

Family	MPS	Per 1			Per 2		
		%	Time	Press	%	Time	Press
F1A	1	1	14.63	PR00			
F3A	1	1	5.73	PR00			
F5A	1	1	13.6	PR00			
F4A	1	1	8.93	PR00			
F9A	1	1	16.2	PR10			
F10A	1	1	9.3	PR00			
F7A	2				1	20.53	PR00
F6A	2	0.32	18.5	PR00	0.68	39.5	PR20
F13A	2				1	47.4	PR10
F15A	2				1	24	PR00
F2A	2				1	13.65	PR00
F8A	2				1	9.96	PR00
F20A	2	0.52	30.5	PR20	0.48	28.3	PR20
F25A	2				1	40.7	PR10
F11A	2				1	34.5	PR00
F12A	2	0.78	44.2	PR00	0.22	12.3	PR00
F28A	2				1	47.2	PR20

Table 8. Percentage of families and respective amount of time processed by period

Columns 3 and 5 indicate the percentage of demand of each family to be processed in each period. Columns 4 and 6 give the amount of processing time in each period. Therefore, family F6A, initially allocated to MPS for processing in the second period, will have 32% of its production, or 18.5 hours, advanced to the first period. F20A and F12A will also have part of their production advanced to the first period.

If the press has not enough capacity to process a family, a heuristic rule that inverts equipment priority, is used. F6A is programmed for the second period, in press PR00. This press does not have enough capacity to process the whole family. Another possible press in this case is PR20, which, during the second period, has the capacity to process only 68% of the family.

Software suggests advancing 32% of the family to the first period, in the prior press - PR00. Table 9 presents the total amount of time spent in each press in both periods. Analysing this table, the manager can decide, for instance, to process F6A in press PR20 in the first period.

Press	∑Times period 1	∑Times period 2
PR00	114.89	1114.94
PR10	16.2	88.1
PR20	30.5	115

Table 9. Work load in presses in two periods

The example presented in this work is the extract of a simulation for 40 families in 4 periods.

In this example some products can be classified as one-stage production scheduling problem in which each activity requires two operations to be processed in a bottleneck stage (resource). A more complex problem is that one in which a product each activity requires two operations to be processed in stages 1 and 2, respectively. In this case there are two options for processing: the first is to produce it by utilizing in-house resources, while the second is to outsource it to a subcontractor. For in-house operations, a schedule is constructed and its performance is measured by the makespan, that is, the latest completion time of operations processed in-house. Operations by subcontractors are instantaneous but require outsourcing cost. Computational model for and approximate algorithm for this NP-hard case can be found in Lee and Choi (2011).

6.2 Auto part gears industry

a. Enterprise characteristics and objectives

A medium-sized manufacturing enterprise with around 200 employers produces auto part gears through 17 cell manufactures numbered from 1 to 25. In general, each cell is dedicated to manufacturing a fixed set of products expressed by a correspondence table, as shown in Figure 15. However, in daily operation there may be a product being processed in another cell, due to a lack of capacity of the priority cell. However, this operation causes disruption in production and should be avoided whenever possible.

The production of gears is determined by the client´s requests. The company works with both regular and special items. For some regular items, the company has a forecast of more than one year, discretized weekly. For others there is forecast of eight months. In the case of special items, the demand is small and irregular. The cells work in two shifts of 8.5 hours per day with a monthly limit of 360 hours'.

Currently the setup time is the time between item entry in the cell and its exit from the last operation. As in the gears process, the setup time of the cells are time consuming (on average 4-5 hours). They are performed concurrently with processing the previous batch, i.e., as the last part of a batch releases one machine of the cell, this machine is immediately prepared for the next batch. In general, this operation is initiated by an operator and completed by the preparer.

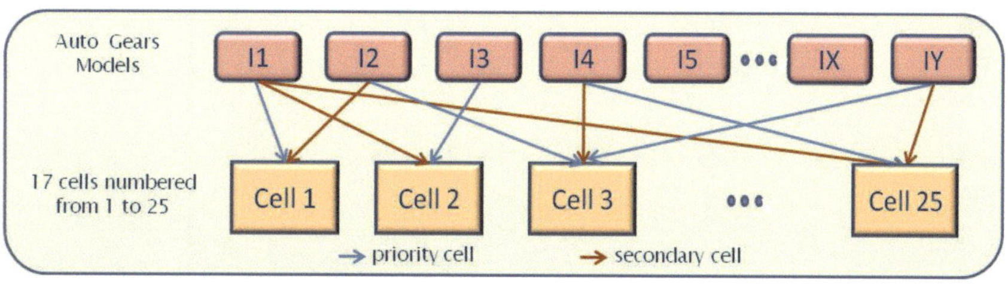

Fig. 15. Products X Cell correspondence

The company has a homemade MRP supported by Excel spreadsheets ® to control shop floor operations. This MRP, as well as in any of the MRP´s encountered on the shelf, does not consider the limits of resource capacity and its method of production programming does not support the decision maker with a temporal production view. In order to support the decision maker in the difficult task of optimally allocating the production lots through the 17 cells along a specified time period, given the quantity demanded and the dates requested, an optimization model based on graph theory was developed. The objective of this graph is to minimize storage and backorders. Taking into account both the availability in the company of the spreadsheet, as well the internal knowledge of this tool, a standard environment for both modelling and presentation of results was developed.

b. Spreadsheet model

As it is a medium-sized enterprise, an excel worksheet was suggested as an environment for the solution, as pointed out in the table 10 whose main function is:

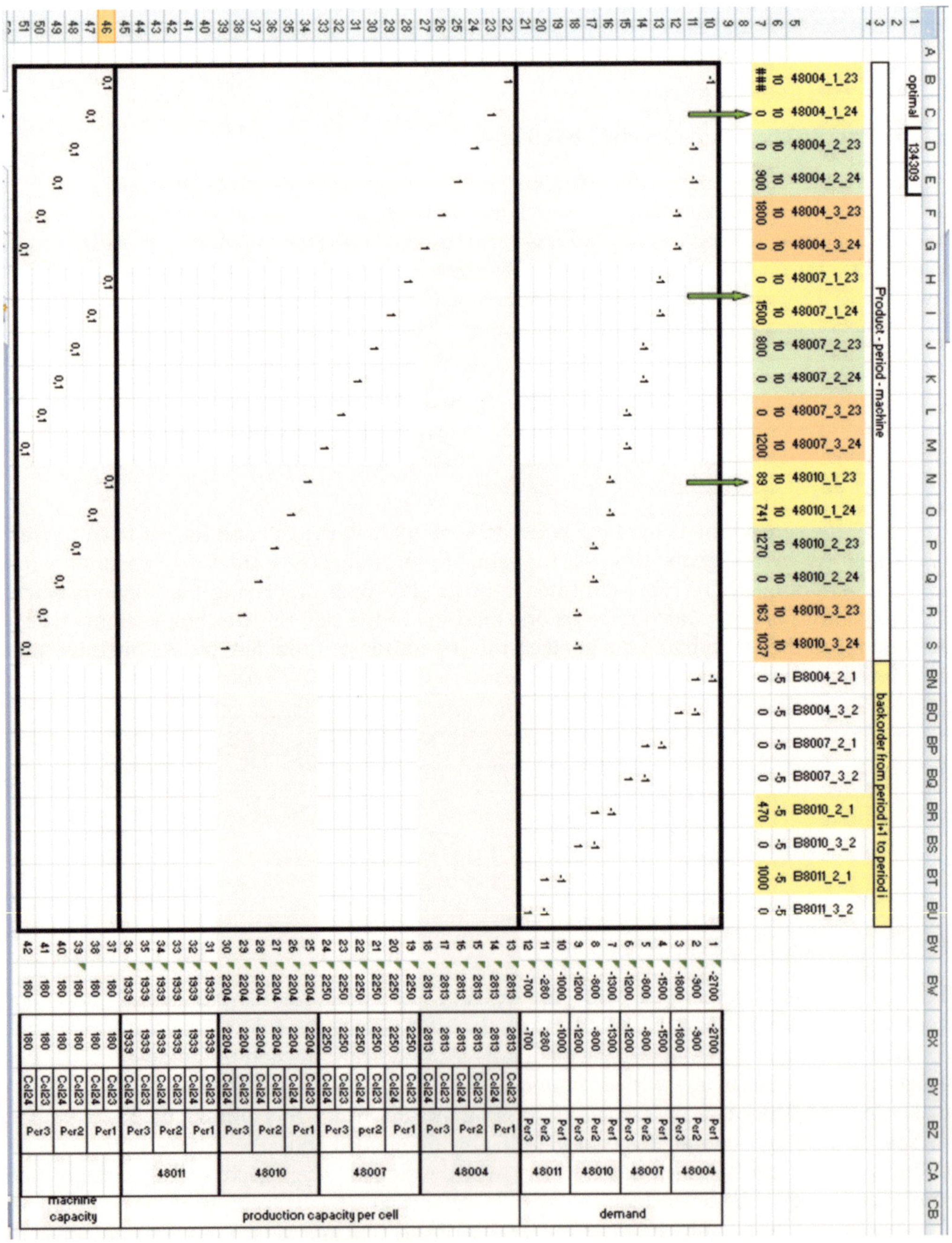

optimal: 134309

Product - period - machine / backorder from period i+1 to period i

Variable	Coefficient	Value
48004_1_23	10	###
48004_1_24	10	0
48004_2_23	10	0
48004_2_24	10	900
48004_3_23	10	1800
48004_3_24	10	0
48007_1_23	10	0
48007_1_24	10	1500
48007_2_23	10	800
48007_2_24	10	0
48007_3_23	10	0
48007_3_24	10	1200
48010_1_23	10	89
48010_1_24	10	741
48010_2_23	10	1270
48010_2_24	10	0
48010_3_23	10	163
48010_3_24	10	1037
B8004_2_1	-5	0
B8004_3_2	-5	0
B8007_2_1	-5	0
B8007_3_2	-5	0
B8010_2_1	-5	470
B8010_3_2	-5	0
B8011_2_1	-5	1000
B8011_3_2	-5	0

No.	Value	RHS	Cell	Period	Product	Constraint
1	-2700	-2700		Per1	48004	demand
2	-300	-300		Per2	48004	demand
3	-1800	-1800		Per3	48004	demand
4	-1500	-1500		Per1	48007	demand
5	-800	-800		Per2	48007	demand
6	-1200	-1200		Per3	48007	demand
7	-1300	-1300		Per1	48010	demand
8	-800	-800		Per2	48010	demand
9	-1200	-1200		Per3	48010	demand
10	-1000	-1000		Per1	48011	demand
11	-280	-280		Per2	48011	demand
12	-700	-700		Per3	48011	demand
13	2813	2813	Cel23	Per1	48004	production capacity per cell
14	2813	2813	Cel24	Per1	48004	production capacity per cell
15	2813	2813	Cel23	Per2	48004	production capacity per cell
16	2813	2813	Cel24	Per2	48004	production capacity per cell
17	2813	2813	Cel23	Per3	48004	production capacity per cell
18	2813	2813	Cel24	Per3	48004	production capacity per cell
19	2250	2250	Cel23	Per1	48007	production capacity per cell
20	2250	2250	Cel24	Per1	48007	production capacity per cell
21	2250	2250	Cel23	per2	48007	production capacity per cell
22	2250	2250	Cel24	per2	48007	production capacity per cell
23	2250	2250	Cel23	Per3	48007	production capacity per cell
24	2250	2250	Cel24	Per3	48007	production capacity per cell
25	2204	2204	Cel23	Per1	48010	production capacity per cell
26	2204	2204	Cel24	Per1	48010	production capacity per cell
27	2204	2204	Cel23	Per2	48010	production capacity per cell
28	2204	2204	Cel24	Per2	48010	production capacity per cell
29	2204	2204	Cel23	Per3	48010	production capacity per cell
30	2204	2204	Cel24	Per3	48010	production capacity per cell
31	1939	1939	Cel23	Per1	48011	production capacity per cell
32	1939	1939	Cel24	Per1	48011	production capacity per cell
33	1939	1939	Cel23	Per2	48011	production capacity per cell
34	1939	1939	Cel24	Per2	48011	production capacity per cell
35	1939	1939	Cel23	Per3	48011	production capacity per cell
36	1939	1939	Cel24	Per3	48011	production capacity per cell
37	180	180	Cel23	Per1		machine capacity
38	180	180	Cel24	Per1		machine capacity
39	180	180	Cel23	Per2		machine capacity
40	180	180	Cel24	Per2		machine capacity
41	180	180	Cel23	Per3		machine capacity
42	180	180	Cel24	Per3		machine capacity

Table 10. Auto part gears spreadsheet model.

Constraints:	=SOMARPRODUTO($B$7:$BU$7;B48:BU48)
Variable cells	$B$7:$BU$7
Restrictions	$B$7:$BU$7
	$B$7:$BU$7
	$B$7:$BU$7

The objective of this implementation is to show how simple it is to develop a model and to develop a schedule using a spreadsheet. The results table shows the scheduling for each product and cell at each time period. It also presents the backorders of 470 units and 1000 units for the products P48010 and P48011 respectively in the period 2.

7. Conclusion

Production scheduling involves operation propositions for real complex systems and despite the ultimate objective of producing a schedule that satisfies domain constraints and optimizes overall performance, scheduling, in the most practical domains, is concerned with solving a problem of a much larger scope involving specification, negotiation and refinement of input requirements and system capabilities. Production scheduling implies the development of a set of activities that lead to most effective overall system performance through simultaneously balancing resource availability, enterprise interests, suppliers constrains and contracts and client needs.

It is therefore unreasonable to expect to fully automate the analysis process in a single system that covers all real applications. The search space is unwieldy and ill-structured, and human expertise is needed to effectively direct the search process. At the same time, problem scale generally demands substantial automation. The research challenge is to offer the user a flexible environment and to incorporate it into the scheduling process, without requiring understanding of the system's internal model. In other words, the system must bear the burden of translating to and from user interpretable representations. It is important to convey the results in a form that facilitates comprehension and conveys critical tradeoffs, and to accept user guidance on how the system model can be manipulated.

Taking into account these aspects, this work used three different approaches to solve real problems: Mathematical Programming (Linear Program), Industrial Practice (ERP) and Manager Expertise, translated into heuristic rules. Mathematical Program results will give a "near optimum" value, since calculations are made based on estimated set-up times aggregated to the processing times. Better representation of set-up times is reached through the MRP module. Knowledge of shop floor particularities allows the manager to identify characteristics that improve the performance of the production system and that can enrich the solution algorithm. The production optimisation balances the capacity with temporal demand oscillation and induces, when necessary and possible, transferring production from overloaded to idle periods. This in turn avoids extra hours and contracting outsider services and to reach the desired production level by advancing or postponing production to idle periods.

The advantage of this integration is clear. It helps the manager deal with multiple approaches added to enterprise competences to support his decisions overcoming the usual disruptions by radical changes in enterprise procedures.

Although this chapter focused on the creation of mathematical models to be coupled with traditional ERP packages there are however numerous alternative model forms and Pidd´s work provides considerations of a wider view of modelling (Pidd, 2003).

New models and methods have to be coded for each new real problem. One way to achieve this is the use of modelling language that provides algebraic and set notations to concisely express mathematical problems that can then be solved using state-of-the-art solvers. These modelling languages, namely AMPL (Algebraic Modelling Language for Mathematical Programming), do not require specific programming skills and provide flexibility to represent production planning models to any enterprise.

8. References

Adler, I.; Resende, M. G. C.; Veiga, G.; Karmakar, N. (1989), An Implementation of Karmakar Algorithm for Linear Programming; *Mathematical Programming*, Vol. 44; No 3, pp 297-335, ISSN: 0025-5610

Al Mutairi, Eid. M. (August 2008). Simultaneous Design, Scheduling and Operation Through Process Integration: A Dissertation Submitted to the Office of Graduate Studies of Texas A&M University in partial fulfillment of the requirements for the degree of DOCTOR OF PHILOSPHY, (06/06/2011), Available from: < http://repository.tamu.edu/bitstream/handle/1969.1/ETD-TAMU-2897/AL-MUTAIRI-DISSERTATION.pdf?sequence=1 >

Bitran, G. R.; Haas, E. A.; Hax, A. C. (1981). Hierarquical Production Planning: A Single Stage System. *Operations Research*, Vol. 29, No 4, (July-August, 1981), pp 717-743, ISSN:0030-364X

Bitran, G. R.; Haas, E. A.; Hax, A. C. (1982). Hierarquical Production Planning: A Two Stage System. *Operations Research*, Vol. 30, No 2, (March-April, 1982), pp 232-251, ISSN:0030-364X

Bowman, E.H.: (1956) Production Scheduling by the Transportation Method of Linear Programming, *Operations Research*, Vol. 4, No. 1, pp. 100-103. ISSN 0030-364X.

Caravilla, M. A.; Souza, J. P. (1995) Hierarchical Production Planning in a Make to Order Company: A Case Study. *European Journal of Operational Research* Vol. 86, No 1, pp. 43-56 ISSN: 0377-2217

Carvalho, M.F; Silva Filho, O.S. (1996), Two-Stage Strategic Manufacturing Planning System: a Practical View, *Proceedings of Twelfth International conference on CAD/CAM Robotics and Factories of the Future*, Middlesex University, London, England, August 1996, ISBN 189825303X.

Carvalho, M.F.; Fernandes, C.A.O; Ferreira, P.A.V.(1999), Multiproduct Multistage Production Scheduling (MMPS) for Manufacturing Systems, *Production Planning & Control*, Vol. 10 No 7, 671-681 ISSN:1366-5871

Clark, A. R., (2003), Optimization Aproximations for Capacity Constrained Material Requirements Planning, *International Journal of Production Economics*, Vol. 84, No 2, pp 115-131, ISSN: 0925-5273

Gershwin, S. B.; Hildebrandt, R. R.; Suri, R.; Mitter, S. K. (1986). A Control Perspective on Recent Trends in Manufacturing Systems; *IEEE Control System Magazine*, Vol. 6 No 2, 3-15 ISSN 0272-1708

Goldratt, E. M and Cox, J. (2004). *The Goal: A Process of Ongoing Improvement* Revised Edition. North River Press, Croton-on-Hudson, NY, ISBN: 9780566086656.

Gondzio, 1996 Multiple centrality corrections in a primal dual method for linear programming, Computational Optimization and application, 6, 137-156, ISSN: 0926-6003.

Graves, C.S.: (1981) A Review of Production Scheduling, *Operations Research*, Vol. 29, No. 4, Operations Management, pp. 646-675, ISSN: 0030-364X.

Hill, Terry. (1991). *Production / Operations Management Text and Cases*, 2nd edition, Prentice Hall, New York. pp 194, ISBN 1435-246X.

Jain A.S; Meeran, S. (1999). A State-of-the-art Review of Job-shop Scheduling Techniques, *European Journal of Operations Research*, Vol. 113 No. 2, pp. 390-434, ISSN: 0377-2217.

Józefowska, J., Zimniak, A.(2008) Optimization tool for short-term production planning and scheduling, Int. *J. Production Economics*, 112, 109–120, ISSN: 0925-5273.

Karimi, B.; Fatemi Ghomi, S. M. T.; Wilson, J. M. (2003). The Capacitated Lot Size Problem: A Review of Models and Algorithms, *OMEGA: The International Journal of Management Science*, Vol. 31, No. 5, (October, 2003), pp 365,378 ISSN: 0305-0483

Kennington, J. (1980) Algorithms for Networking Programming, John Wiley & Sons, pp 166-171, ISBN-10: 047106016X.

Lee, K; Choi, B: (2011) Two-stage production scheduling with an outsourcing option, *European Journal of Operational Research* 213, 489–497, ISSN: 0377-2217

Özdamar, L.; Bozyel, M. A.; Birbil, S. I. (1998). A Hierarchical Decision Support System for Production Planning (with case study). *European Journal of Operational Research*, Vol.104, No. 3, (February, 1988), pp. 403-422, ISSN: 0377-2217.

Pidd M, 2003, *Tools for Thinking: Modelling in Management Science*, John Wiley and Sons Ltd, Chichester, ISBN: 0-470-84795-6

Rodammer, F.; White, P. K.:(1988) A Recent Survey of Production Scheduling, *IEEE Transactions on System, Man and Cybernetics*, vol. 18, No. 6, 841-851, ISSN: 0018-9472.

Smith, Stephen F. (2003). Is Scheduling a Solved Problem?, *Proceedings of First Multidisciplinary International Conference on Scheduling: Theory and Applications (MISTA)*, Nottinghan, UK, August 2003, ISBN 0-954-5831-2-8

Voβ, Stefan; Woodruff David L. (2006). *Introduction to Computational Optimization Models for Production Planning in a Supply Chain* (2nd edition), Springer, Germany, pages 16, 17, 29, 31, 48, 202, ISBN: 3540298789.

Wang, G; Gong, W; and Ryan Kastner, R. (2008) *Operations Scheduling: Algorithms and applications* in High-Level Synthesis, chapter 13 Springer Science + Business Media B.V., ISBN 978-1-4020-8587-1.

Wright, S.J. (1996) Primal-Dual Interior Point Methods, *SIAM* Publications, SIAM, Philadelphia, PA, USA.

Wu, T.; Shi, L.; Geunes, J.; Akartunalı, K.: (2011) An optimization framework for solving capacitated multi-level lot-sizing problems with backlogging, *European Journal of Operational Research* 214. 428–441, ISSN: 0377-2217.

Yamakami, A.Takahashi, M.T., Carvalho, M.F. (2000) Comparision of some algorithms for manufacturing production planning, *IFAC-MIM 200 Symposion on Manufacturing, Management and Control*, 280-284.

Yang, L.; Ng, C.T., Cheng, T.C.E. Optimal production strategy under demand fluctuations: Technology versus capacity (2011) European Journal of Operational Research 214 (2011) 393–402. ISSN:0377-2217

Zahorik, A.; Thomas, L. J.; Trigeiro, W. W. (1984). Network Programming Models for Production Scheduling in Multi-Stage, Multi-Item Capacitated Systems. *Management Science*, Vol. 30, No 3, (March, 1984) pp 308-325, ISSN: 0025-1909

Part 5

Design and Implementation of Production Scheduling Systems

9

Achieving Cost Competitiveness with an Agent-Based Integrated Process Planning and Production Scheduling System

Ming Lim[1] and David Zhang[2]
[1]*Aston University,*
[2]*University of Exeter,*
United Kingdom

1. Introduction

As globalisation takes place, the market is getting more and more competitive. Manufacturing enterprises are facing tremendous pressure to succeed in the market with promising market shares. This has led to enterprises seeking for competitive advantage in order to vie with their rivals. In manufacturing context, efforts have been put in, for instance, to reduce production lead times, maximise productivity and optimise resource utilisation. These efforts aim to reduce all types of costs incurred, as a means to achieve cost competitiveness. There is also a challenge for the manufacturers to be competent to efficiently and cost-effectively cope with dynamic changes of customer demands in the market. These demands are related to a wide range of product mixes with short product lifespan and with unpredicted demand pattern (Zhang, 2011).

From a manufacturing perspective, the efficiency of manufacturing operations (such as process planning and production scheduling) are the key element for enhancing manufacturing competence. Process planning and production scheduling functions have been traditionally treated as two separate activities, and have resulted in a range of inefficiencies. These include infeasible process plans, non-available/overloaded resources, high production costs, long production lead times, and so on (Saygin & Kilic, 1999; Khoshnevis & Chen, 1993; Zhang, 1993). Above all, it is unlikely that the dynamic changes can be efficiently dealt with. Despite much research has been conducted to integrate process planning and production scheduling to improve manufacturing efficiency, there is still a gap to achieve the competence required for the current global competitive market.

In this research, the concept of multi-agent system (MAS) is adopted as a means to address the aforementioned gap. A MAS consists of a collection of intelligent autonomous agents able to solve complex problems. These agents possess their individual objectives and interact with each other to fulfil a global goal. This chapter describes a novel use of an autonomous agent system to facilitate the integration of process planning and production scheduling functions to cope with unpredictable demands. This refers to the uncertainties in product mixes and demand patterns. The novelty lies with the currency-based iterative agent bidding mechanism to allow process planning options and production scheduling options to be evaluated simultaneously, so as to search for an optimised and cost-effective solution. This agent based system aims to achieve manufacturing competence by means of enhancing the flexibility and agility of manufacturing enterprises.

This chapter is organised as follows. Section 2 reviews the literature of the existing approaches to integrated process planning and production scheduling. The limitations of these approaches will also be discussed. Section 3 describes the concept of MAS and Section 4 introduces the currency-based iterative agent bidding mechanism proposed in this study. A Tabu search optimisation technique to facilitate the adjustment of current values for agent bidding is presented in Section 5. Section 6 discusses the findings of the simulation results for the iterative bidding mechanism and further analyses the bidding results with three heuristic integrated process planning and scheduling approaches. Section 7 concludes this chapter.

2. Integrated process planning and production scheduling

In the literature, there is a vast number of research works reported contributing to the integration of process planning and production scheduling. These include non-linear process planning, flexible process planning, closed-loop process planning, dynamic process planning, alternative process planning, and just-in-time process planning (Cho & Lazaro, 2010; Kim et al., 2009; Moslehi et al. 2009; Omar & Teo, 2007; Wang et al., 2003; Saygin & Kilic, 1999; Usher & Fernandes, 1996; Khoshnevis & Mei, 1993). According to Larsen & Alting (1990), these works can be classified into three broad categories: non-linear process planning (NLPP), closed-loop process planning (CLPP), and distributed process planning (DTPP).

2.1 Non-Linear Process Planning (NLPP)

NLPP entails a planning system that generates a list of possible alternative plans for each part prior to actual production on the shop floor. This means that NLPP is based on a static shop floor condition. All these possible plans are ranked according to process planning criteria. The first priority plan is always used when the job is required. If the plan is not suitable, e.g. due to resource unavailability, the lower priority plan will be chosen. This procedure is repeated until a suitable plan is found. Examples of such system include FLEXPLAN that uses reactive re-planning strategies to allow fast reaction when unexpected events occur on the shop floor during the execution of a schedule (Tonshoff et al., 1989), and a framework by Hou & Wang (1991) that firstly disaggregating the process planning problems and followed by generating alternative process plans for the parts to be manufactured. Other similar works include Ho & Moodie (1996), Hutchinson & Pflughoeft (1994), and Srihari & Greene (1988).

Recognising the weaknesses of NLPP, some researchers proposed the idea of a two-stage approach to improve NLPP. In the first stage, all possible alternative process plans that do not take into account of operational status of the shop floor resources are generated. The second stage is dynamic process planning whereby the generated process plans are evaluated by taking into account of the availability of the shop floor resources and the objectives or rules are specified by the scheduler. The result of this two-stage approach is a set of ranked near-optimum alternative plans and schedules. The systems applying such an approach are PARIS (Usher & Fernandes, 1996), DYNACAPP (Ssemakula & Wesley, 1994), and THCAPP-G (Wang et al., 1995).

2.2 Closed-Loop Process Planning (CLPP)

NLPP offers flexibility to the scheduling department with a list of alternative process plans. However, process planners do not take into account of the shop floor condition and an arbitrary set is generated based on their experience. In turn, production schedulers only use the alternatives that are available. To make the process planning more efficient, there is a need to have feedback from the shop floor with detailed information of the shop floor condition as well as requirements from scheduling department. With this information, no further effort will be spent on investigating alternatives that are of no use. Furthermore, the risk of overlooking important aspects (e.g. machine reliability and utilisation, bottlenecks) is also reduced. CLPP is an approach that could provide such feedback.

CLPP generates plans for jobs in real time based on the status of the resources at that time. Production schedulers provide process planners with information in relation to resource availability so that every plan is feasible with respect to the current availability of production facilities. Real time status has become a crucial element for CLPP and therefore, CLPP is also referred to as real time process planning or dynamic process planning. The research works based on this approach include a heuristic algorithm proposed by Khoshnevis & Chen (1989) developing a dynamic list of available machines and a list of features for each part. When a match is found between the two lists, the part will be assigned to that machine. However, the algorithm has neglected one issue in relation to the allocation of producing the features to machines. For instance, the algorithm may have allocated a feature to a less desirable machine at a given instant, whereas had it waited for a short while, a more desirable machine might have become available. The authors then introduced the concept of time window into their improved algorithm to deal with this problem (Khoshnevis & Chen, 1990). Although the improved algorithm can yield better results, the computational complexity is increased. In a later work by Chen & Khoshnevis (1992), the integration problem is viewed as a scheduling problem with flexible process plans. The priority is given to the scheduling module. Whenever an assignment of an operation to a machine is made by the scheduling module, the process planning module is invoked to check the validity of the assignment. Other examples of using CLPP are Kiritis & Porchet (1996) and Iwata & Fukuda (1989).

In NLPP, feedback information from the shop floor (i.e. information on the shop floor condition and requirements from scheduling department) is provided to the process planning department and as a result, process planning can be performed more efficiently and infeasible plans (i.e. due to unavailability of resources) can be eliminated. However, the aforementioned manufacturing competence is still not yet achieved in the proposed works. Despite the elimination of infeasible plans, the cost reduction through optimisation of utilisation of resources and minimisation of bottlenecks are not achieved in NLPP.

2.3 Distributed Process Planning (DTPP)

DTPP is a promising approach that performs both process planning and production scheduling simultaneously in a distributed manner, starting from a global level (i.e. pre-planning) and ending at a detailed level (final planning). In DTPP process planning and production scheduling activities are carried out in parallel and in two phases. The first phase is pre-planning whereby process planning function analyses the jobs/operations to be carried out. The features and feature relationships are recognised and the corresponding manufacturing processes are determined. The required machine capabilities are also estimated. The second phase is final planning, which matches the required operations with the operational capabilities of the available manufacturing resources. The integration occurs at the point when resources are available and the operation is required. In this integration, process planning and production scheduling are carried out simultaneously. This approach is sometimes also referred to as just-in-time process planning. The result of this approach is dynamic process and production scheduling constrained by real-time events. Such approach includes the early works by Mamalis et al. (1996), Zhang (1993), Mallur et al. (1992), and more recently by Li et al. (2010), Moon et al. (2009), and Wang et al. (2009).

Despite the effort to integrate process planning and scheduling to find satisfactory solutions, further work is required for the solutions to be optimised in respond to dynamic changes in order to enhance the agility of manufacturing systems. To achieve overall optimality, rescheduling alone may not be effective (e.g. Wang et al., 2011). Process planning options should be taken into consideration to provide flexibility and optional scenarios in using alternative resources to respond constantly to dynamic changes. This means that process planning options and production scheduling options should be integrated and optimised dynamically, so that constraints from both functions can be fulfilled simultaneously and a near-optimum integrated plan and schedule can then be produced. Furthermore, the integration of process planning and production scheduling should also be able to provide scenarios where the production operational structures and possible reconfiguration of manufacturing systems can be assessed. By enhancing this manufacturing competence, the cost competitiveness will be achieved. In this research, a multi-agent system (MAS) is employed aiming to achieve this.

3. Multi-Agent System (MAS)

MAS is a popular research technique applied in various disciplines. A MAS is a distributed intelligent system consisting of a population of agents that pursue individual objectives and interact closely with each other to achieve a global goal. Each agent represents an entity (e.g. a machine or a job) and is endowed with a certain degree of autonomy and intelligence, which includes the ability to perceive its environment and to make decisions based on its knowledge (Ferber, 1999).

In a MAS, a complex system is decomposed into autonomous and loosely-coupled subsystems represented by agents (Wooldridge, 1997). The term autonomous refers to the independency of control between the agents. Each agent determines its course of actions and other agents may influence an agent's decision by means of coordination (through collaboration or competition/negotiation). The term loosely coupled refers to the dependency on information between the agents. This dependency may exist for some tasks and shall not oblige to overload one agent's capability. Agents that represent the subsystems are able to solve problems in their domain with their own thread of control and execution. They carry out tasks autonomously without depending on other agents. The agent characteristics of intelligence and autonomous decision-making architecture have attracted many researchers in manufacturing domain solving complex manufacturing problems, including research related to process planning and production scheduling.

In general, the agent-based process planning and production scheduling approaches found in the literature can be grouped into two categorises based on the interaction mechanism used by the agents. They are bidding based methods and non-bidding based methods. Bidding based methods include the works by Robu et al. (2011), Kumar et al. (2008), Liu et al. (2007), Lima et al. (2006), Wong et al. (2006), and non-bidding works by Hajizadeh et al. (2011), Blum & Sampels (2004), Caridi & Sianesi (2000), and Ottaway & Burns (2000).

For any MAS, the design of agents is crucial to ensure the global goal and individual objectives are both fulfilled. This includes the agent functions and network, agent interaction mechanism and its protocols for coordination. In this chapter, the authors proposed a novel use of an autonomous agent system to facilitate the integration of process planning and production scheduling functions in order to maximise the manufacturing competence to cope with unpredictable demands. The novelty lies with the currency-based iterative agent bidding mechanism to allow process planning options and production scheduling options to be evaluated simultaneously, so as to search for an optimised, cost-effective solution. This agent based system aims to provide the flexibility and agility of manufacturing enterprises required to cope with the uncertainties in the market. The following section discusses this iterative agent bidding mechanism in detail.

4. Iterative agent bidding mechanism

In the proposed iterative agent bidding mechanism, a currency-like metric is used whereby each operation to be performed is assigned with a virtual currency value. These operations will then be announced to the agents (e.g. representing resources on the shop floor) and they will bid for the operations based on the currency values. These currency values are used as a parameter to control the bidding process between agents. Agents will only put forward the bids for the operation if they make a virtual profit (i.e. the difference between the given currency for the operation and the cost of performing the operation) that is above a virtual profit threshold set by that agent. This means that these parameters have a direct influence over the decisions of agent bidding for operations and forwarding the bids; therefore the adjustment of parameters will result in different bids constructed. In this mechanism, the virtual currency values will be adjusted iteratively, and so does the bidding process between agents based on the new set of currency values generated. This is to search for better and better bids, leading to near-optimality. The iterative bidding mechanism aims to achieve the lowest possible total production cost while satisfying the delivery due dates. Moreover, with the adjustment of currency values it is able to drive the behaviour of agents in a way that agents become proactive if they know they can perform the job with greater amount of virtual profit earned and vice versa.

The iterative bidding mechanism is illustrated in Figure 1. Assume that machine agents representing the machines on the production shop floor and a job agent representing a job (e.g. to produce a component) to be performed which can be broken down into a number of operations (e.g. to produce the features of the component). The iterative bidding mechanism takes place between the job agent and machine agents. As depicted in Figure 1, the bidding process begins when the job agent announces the job to be performed to all machine agents to bid (*Step 1*). The announcement includes information related to the machining operations to be carried out, such as the number and type of machining operations, recommended type of machining processes for the operations, etc., and the virtual currency value assigned to each operation. Machine agents that are able to perform the first operation will come forward to become 'leaders' whose responsibility is to group other machine agents to perform the remaining operations (*Steps 2-3*). The number of leaders indicates the number of virtual machine groups.

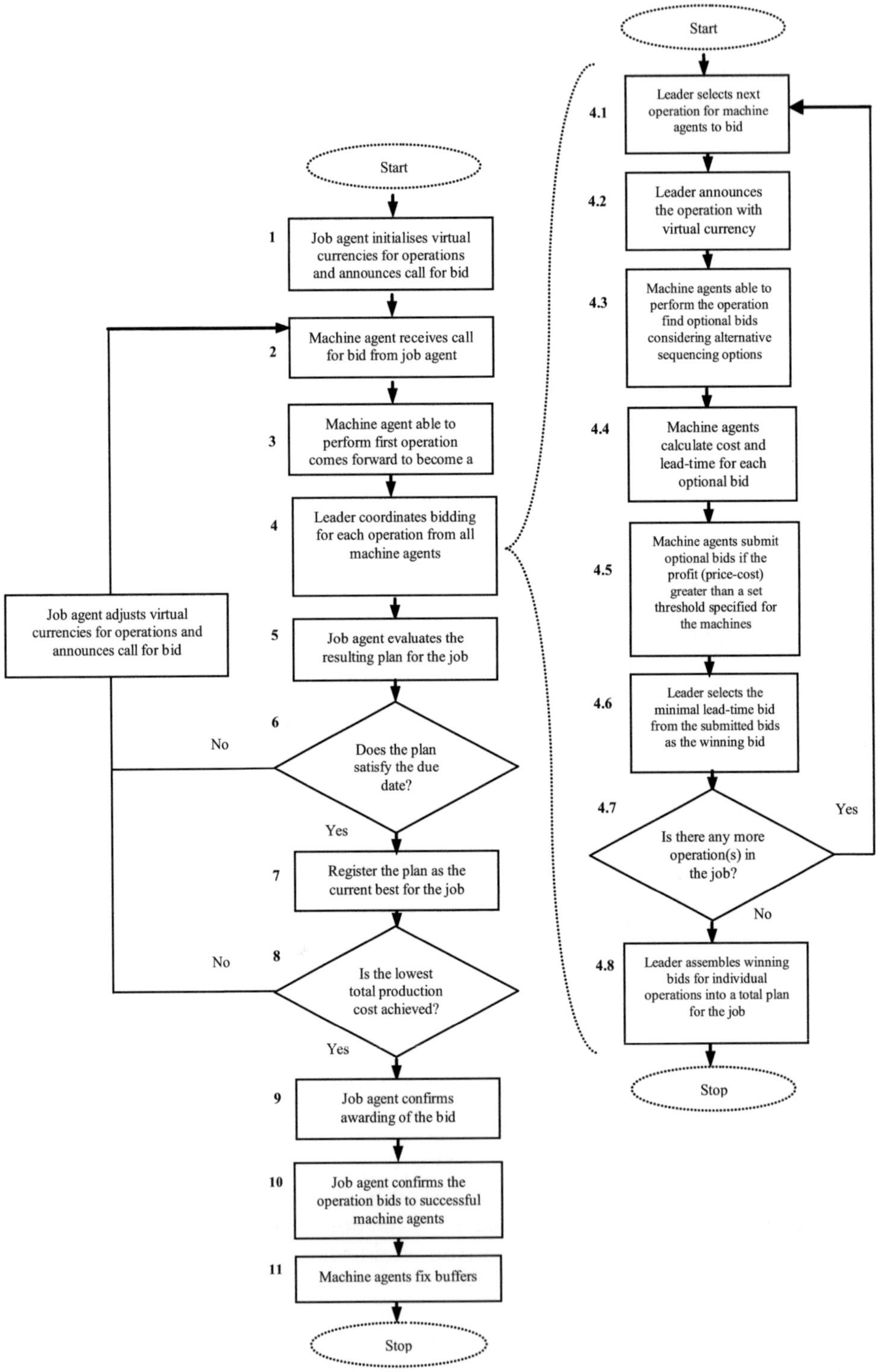

Fig. 1. Iterative agent bidding mechanism

After the leaders are selected, they announce the second operation to all machine agents, including the leaders themselves (*Step 4*). Machine agents that are able to carry out the operation will come forward to bid for the job. They may reschedule and optimise their job buffer by shifting jobs if other operations' due dates are not violated. This aims to produce optional and better bids *BOi,x,k* (*BOi,x,k* denotes the kth bid option from machine *rx* for operation *Oi*). In this way, bottlenecks can be reduced and machine utilisation can also be optimised. By shifting jobs in the job buffer, some bids may eliminate tool change and setup activities and as a result, the time needed to carry out the operation could be reduced. However, extra cost might be involved due to the job shifting in the job buffer, e.g. holding for work-in-process. Machine agents work out their bids in terms of production cost and lead time. The individual machine production cost is obtained as:

$$C_i = C_{ti} + C_{wi} + C_{si} + C_{pi} + C_{ri} \tag{1}$$

where

$$C_{ti} = C_{ti/d}(D) \tag{2}$$

$$C_{pi} = C_{pi/t}\left(\frac{V_{removed}}{MRR}\right) \tag{3}$$

where

C_{ti} = transportation cost from the location of preceding machine (unit of cost),
$C_{ti/d}$ = transportation cost / unit of distance (unit of cost),
D = distance from the location of preceding machine (m),
C_{wi} = holding cost (unit of cost),
C_{si} = setup cost (unit of cost),
$V_{removed}$ = volume to be removed in order to produce the feature (mm3),
MRR = material removal rate (mm3 / unit of time),
C_{pi} = processing cost (unit of cost),
$C_{pi/t}$ = processing cost / unit of time (unit of cost),
C_{ri} = rescheduling cost (unit of cost).

The machine production cost function used in this study does not, however, truly reflect the actual production cost in real production. The cost function is developed for evaluation purposes and the costs such as material cost and labour cost are disregarded.

The individual lead time is worked out as:

$$T_i = T_{ti} + T_{wi} + T_{si} + T_{pi} \tag{4}$$

where

$$T_{ti} = T_{ti/d}(D) \tag{5}$$

$$T_{wi} = \sum_{j=1}^{n} t_{wi[j]} \tag{6}$$

$$T_{pi} = \frac{V_{removed}}{MRR} \tag{7}$$

where

T_{ti} = transportation lead time from the preceding machine (unit of time),

D = distance from the location of preceding machine (m),

$T_{ti/d}$ = transportation lead time / unit of distance (unit of cost),

T_{wi} = waiting time at buffer, i.e. queuing time or bottlenecks (unit of time),

$\sum_{j=1}^{n} t_{wi[j]}$ = total waiting time of n jobs scheduled in the job buffer before the currently bidding job (unit of time),

T_{si} = setup time (unit of time),

T_{pi} = processing lead time (unit of time)

$V_{removed}$ = volume to be removed in order to produce the feature (mm^3), and

MRR = material removal rate (mm^3 / unit of time).

Each machine agent decides whether to forward a bid based on the amount of virtual profit earned:

$$P_{i,x,k} = CU_i - C_{i,x,k} \tag{8}$$

where $P_{i,x,k}$ is the virtual profit that could be made by machine r_x on operation O_i with bid option $BO_{i,x,k}$, CU_i is currency value assigned to operation O_i , and $C_{i,x,k}$ is the production cost for r_x to carry out O_i with bid $BO_{i,x,k}$ as defined by Eq. 1. If $P_{i,x,k}$ is above a set threshold P_{tx} (i.e. $P_{i,x,k} \geq P_{tx}$), the bid will be put forward to the leader. P_{tx} is a mark-up profit that is based on the production cost $C_{i,x,k}$ i.e. $P_{tx} = C_{i,x,k} + C_{i,x,k} \,.\, M_{i,x,k}$., where $M_{i,x,k}$. is a random value in the range [0,N], and N is a limiting percentage value. By shifting jobs in the job buffer, a machine agent may put forward more than one bid as long as the virtual profits of the bids are above the set threshold. The threshold varies from one machine to another based on the cost of machine. However, if the profit is below the set threshold ($P_{i,x,k} < P_{tx}$), the machine agent will not forward the bid to the leader. In mathematical terms:

$$B_{i,l} = BO_{i,x,k},\ T_i^{(l)} = T_{i,x,k},\ C_i^{(l)} = C_{i,x,k} \quad \text{if } P_{i,x,k} \geq P_{tx} \tag{9}$$

where $B_{i,l}$ denotes the lth bid submitted for operation O_i, $T_{i,x,k}$ is the lead time for r_x to carry out O_i with bid option $BO_{i,x,k}$, $T_i^{(l)}$ and $C_i^{(l)}$ are the lead time and cost for carrying out O_i with bid $B_{i,l}$. When the bids are received, the leader selects the best bid that provides the shortest lead time from all bids put forward by machine agents:

$$B_i^{win} = B_{i,l},\ T_i^{win} = T_i^{(l)},\ C_i^{win} = C_i^{(l)}$$
$$if\ \ T_i^{(l)} = \min\left(T_i^{(1)}, T_i^{(2)}, \ldots\ldots, T_i^{(L)}\right) \tag{10}$$

where B_i^{win} represents the winning bid for O_i, T_i^{win} and C_i^{win} are the lead time and cost corresponding to the winning bid, L is the total number of bids submitted for O_i.

The bid messages can be used to reflect a variety of dynamic status information (e.g. machine status, order condition), and therefore making the bidding mechanism suitable for real-time operational controls. This grouping process continues until all the operations in the job have been scheduled to the most appropriate machines. When the leaders have virtually grouped other machines to perform all operations *(O1, O2, ..., On)*, they put together all the individual production costs (i.e. total production cost) and lead times (i.e. total lead time) of the selected machines, and forward the one complete bid as a machine group to the job agent for evaluation (*Step 5*). This bid consists of the total lead time and total production cost denoted as follow:

$$T = \sum_{i=1}^{n} T_i^{win},\ \ C = \sum_{i=1}^{n} C_i^{win} \tag{11}$$

The job agent evaluates the bids by means of satisfying the due date D at minimal total production cost

$$Min\left(C = \sum_{i=1}^{n} C_i\right)$$
$$T = \sum_{i=1}^{n} T_i \le D \tag{12}$$

If the due date is not satisfied (i.e. $T > D$), the virtual currency allocated to operations will be adjusted in the next iteration to look for a better plan (*Steps 6-8*). The lead time and cost of a plan resulting from a bidding iteration are dependent on the virtual currencies. Higher virtual currencies for operations increase the attractiveness of the operations to machine agents and encourage the agents to submit more bids for the operations (even though some bids may bear higher costs) and vice-versa. The iterative loop stops when a near-optimum plan that satisfies the due date with considered near-minimum cost is found. When the near-optimum plan is obtained, the job agent will award the job to the machine group that meets the due date and provides the minimum total production cost. The machine agents in the awarded machine group will then commit to the operations awarded by updating their loading schedules (*Steps 9-11*). If the product orders are large and consistent, there could be a need to group the machines in this virtual machine group physically (i.e. reconfiguring the layout of the existing manufacturing system), which may improve the system, as well as cost efficiency. In this way, the reconfiguration of manufacturing systems can be assessed.

Each agent has individual objectives and a global goal to achieve. For this proposed MAS, the global goal is to find an optimised process plan and schedule that gives the lowest production cost while satisfying all requirements such as due date and product quality. As for individual objectives, the machine agents strive to give the best performance in order to win the jobs and optimise its machine utilisation, and the job agent is responsible for assigning the operations to the outstanding group of machines. Via the iterative bidding mechanism and bid evaluation, agents with different objectives will come to a point where the agents' objectives and the global goal can be satisfied.

A Tabu search optimisation technique is employed in this study to investigate how and to what degree the currency values should be adjusted in each iteration in order to obtain better solutions (leading to near-optimality) for integrated process planning and scheduling problems. This technique will be discussed in the following section.

5. Tabu search optimisation technique

The basic form of TS approach was founded on the ideas proposed by Glover (1989, 1990). This approach is based on the procedures designed to cross boundaries of feasibility or local optimality, which are usually treated as barriers. In other words, TS is a meta-heuristic that guides a local heuristic search procedure to explore the solution space beyond local optimality. The key parameters used in TS are as follows:

- Tabu move – a move that is forbidden because it has previously been taken in the search process.
- Aspiration criterion – a criterion to remove the Tabu move that considered to be sufficiently attractive leading to a better solution.
- Intensification strategy (a.k.a. short-term memory) – a rule that encourages moves surrounding the solution that previously found good.
- Diversification strategy (long-term memory) – a rule that encourages the search process to examine unvisited regions and to generate solutions that are difference from those visited before.

With the simplicity of applying the concept of TS, many researchers have adopted TS in production research for optimisation purposes. Baykasoglu & Ozbakir (2009) proposed a multiple objective TS framework to generate flexible job shop scheduling problems with alternative process plans in order to analyse its performance and efficiency. Demir et al. (2011) proposed a TS approach to optimise production buffer allocation in order to enhance manufacturing efficiency. Baykasoglu & Gocken (2010) used TS to solve fuzzy multi-objective aggregate production planning system. Xu et al. (2010) used a two-layer TS approach to schedule jobs with controllable processing times on a single machine in order to meet costumer due dates.

The TS approach proposed in this study is described in Figure 2. With an illustration of a component that has five features, the approach started off with initialising all the relevant parameters such as initial solution (i.e. a set of currency values), Tabu list size for intensification and diversification strategies, and stopping criteria. This approach consists of two main operators or moves, i.e. intensification (currency values adjustment) and diversification (pairwise exchange). For intensification, every currency value has an equal opportunity to be selected for currency adjustment. If a move j is tabu-active (i.e. Tabu $[j] \neq 0$), it is not supposed to be chosen again. However, an aspiration criterion can be applied in the case if the tabu-active move j creates a better solution (i.e. lower cost) than the overall best solution found so far. At each move, the solutions generated will be evaluated and compared to the overall best solution and subsequently the overall best solution will be updated if the new solution found outperforms the overall best solution. Eventually, after the pre-determined number of intensifications to be carried out (M) or the number of moves when no consecutive improvement was found (K) is reached, diversification will take place to explore new regions (i.e. pairwise exchange of currency in the initial solution X_0). This process continues until all the regions are explored. The simulation results of this approach will be discussed in next section.

Step 1: *Initialise the following parameters*

- Initial solution, X_0 (CU_1 CU_2 CU_3 CU_4 CU_5) and objective function, $F(x_0)$.
- Best solution, $X^* = X_0$ and best objective function, $F^* = F(x_0)$.
- Tabu list size t for Intensification tabu list.
- Number of intensifications to be carried out, M.
- Counter for diversification, n.
- Diversification size that depends on the array size (stopping criterion), N.
- Counter for no consecutive improvement was made, k.
- Size of no consecutive improvement was made, K.

Intensification Tabu List (currency values alteration)

(short-term memory)

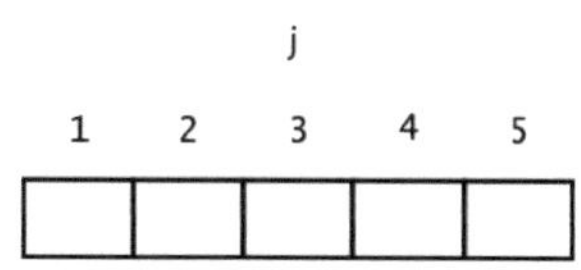

A currency will be randomly selected to be decreased/increased by α%. Once a currency is selected, it will be marked with t (i.e. Tabu-active for t iterations). Tabu [j] = t, where t is a scalar, representing Tabu list size. The value of t is reduced by 1 at every iteration.

Diversification Tabu List (pairwise exchange)

(Long-term memory)

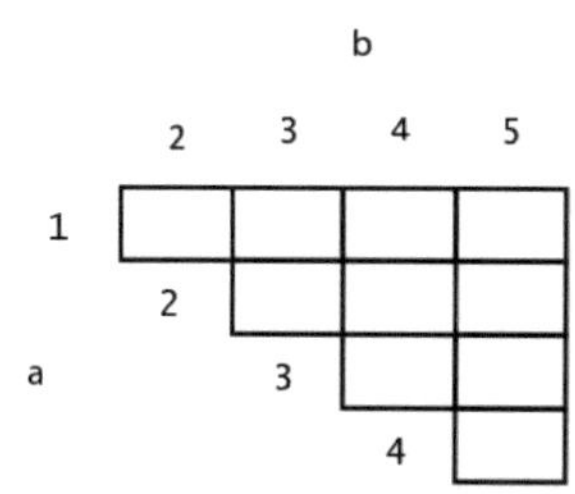

Tabu [a][b] (for $a < b$) stores numerical value in the ath row and b^{th} column of the array. When a exchanges with b, Tabu [a][b] will be marked. The pairwise exchange is in random order. This move is in the tabu list, which means no such move is permitted in the whole searching process. This is used to diversify the search into new regions. The search process will terminate when all Tabu [a][b] have been marked.

Step 2: Iteration $i = 0,1,2,\ldots M$. X_i denotes the current solution.

Fig. 2. Tabu search approach for iterative agent bidding mechanism

6. Test case and results analysis

The proposed MAS was implemented on a Java platform; a test case was used to simulate the effectiveness of the currency-based iterative bidding mechanism. In the test case, 10 machines (4 lathe machines, 3 milling machines and 3 drilling machines) were operating on the shop floor, and each machine has different capacity and capability. The machines data is depicted in Table 1. These machines are served by automated guided vehicles (AGVs) and each machine has its own buffer of jobs (of components C1, C2, C3 and C4) that have been previously scheduled (Table 2).

Machine	*Process*	*Co-ordinate X (location)*	*Co-ordinate Y (location)*	**Reliability**	**Setup time**	**Setup Cost**	**Misc Cost**	**Machining Cost**	**Holding Cost**	**MRR**
Lathe L1	Turning Drilling	40	0	0.8	25	2.5	160	1.6	0.25	36
Lathe L2	Turning Drilling	80	0	0.75	30	2.6	160	2.0	0.25	40
Lathe L3	Turning Drilling	120	0	0.65	35	2.9	170	2.1	0.20	30
Lathe L4	Turning Drilling	160	0	0.9	28	2.6	160	1.5	0.25	29
Mill M1	Milling	100	40	0.95	20	2.4	180	1.2	0.25	32
Mill M2	Milling	100	80	0.75	22	2.6	180	1.4	0.25	32
Mill M3	Milling	100	120	0.85	24	2.5	200	1.5	1.5	36
Drill D1	Drilling	0	40	0.75	32	3	200	1.8	0.4	37
Drill D2	Drilling	0	80	0.9	28	2.8	190	1.9	0.35	40
Drill D3	Drilling	0	120	0.85	29	2.8	200	1.5	0.25	30

Table 1. Machines data

Lathe	Schedule	Milling	Schedule	Drilling	Schedule
L1	C4(2)* 620-830^	M1	C3(2) 500-792 C3(3) 792-988	D1	C1(1) 0-240
L2	C1(2) 280-620	M2	C2(1) 0-380 C4(3) 1000-1179	D2	C2(2) 460-650
L3	C3(1) 0-415 C2(3) 750-925	M3	C4(1) 0-432 C1(3) 670-878	D3	C2(4) 1200-1380
L4	C1(4) 940-1265				

*C*x*(*y*) means job sequence *y* of component *x*
^ in unit of time

Table 2. Machines job schedule

To evaluate the bidding mechanism, this test case consists of three components of which orders were placed at interval times. These components are ComA, ComB and ComC. Table 3 listed the process sequence of producing the features of the components, and the information related to the currency values, removal volumes, and tolerance requirements of each feature in the components.

The simulation process begins with the job agent announcing the jobs of producing ComA to all the machine agents. This process repeats for ComB and ComC. To discuss the implementation in details, the simulation process for ComC is predominantly discussed in this section. The assumptions made in the implementation are:

- A machine only performs one process at a time
- A component can be machined by the same machine more than once
- All the machines are accessible by AGVs
- The AGVs are considered to be always available
- Each machine has infinite capacity input and output buffers
- Auxiliary processes for surface treatment such as grinding and reaming are not considered
- Material and labour costs are disregarded
- Chip formation, cutting fluids, temperature rise, and tool wear due to cutting process are neglected.

During the simulation, two test runs are carried out with the value of α (for the adjustment of currency values in intensification process) set at 15% and 30% respectively. In both test runs, the number of moves for intensification process is set to 10 and the Tabu list is 1. In the first test run, the simulation completed at 110th moves and the near-optimum bid obtained has a production cost of 3224 units and a lead time of 1374 units. Figure 3 illustrates all the bids obtained at each TS move.

Com ID	Quantity	Due date (units of time)	Features to produce in sequence	Process required	Currency value	Removal volume (cm^3)	Tolerance (+/- mm)
ComA	50	900	Hole (Blind, flat-bottomed)	Drilling	575	55	0.75
			Hollow Cylinder (Through)	Turning	770	140	1.25
			Slot	Milling	450	30	1.00
			Hole (Centre, blind, flat-bottomed)	Drilling	500	40	075
ComB	45	1000	Slot	Milling	780	180	1.25
			Hole (Centre, blind, flat-bottomed)	Drilling	460	30	1.00
			Slot	Milling	440	60	1.25
ComC	70	1400	Hollow Cylinder (Through)	Turning	850	150	1.25
			Hollow Cylinder (Through)	Turning	425	40	1.25
			Hollow Cylinder (Through)	Turning	600	70	1.25
			Slot	Milling	380	30	1.00
			Hole (Blind, flat-bottomed)	Drilling	600	50	0.75

Table 3. New components to produce

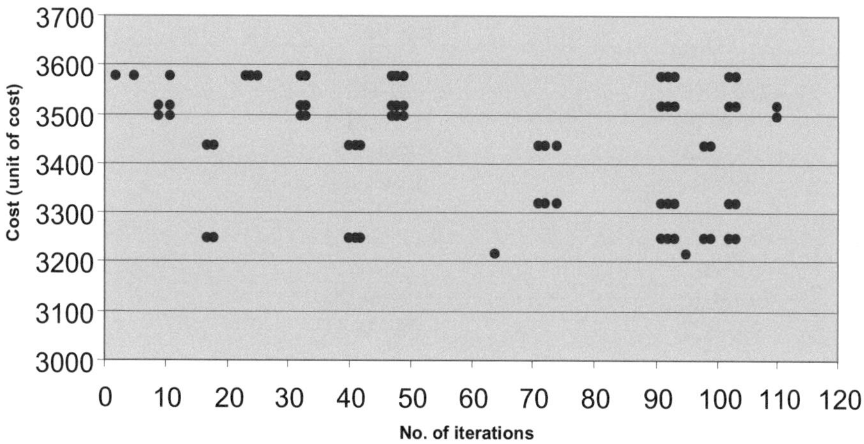

Fig. 3. Bids received at each TS move ($\alpha = 15\%$).

In Figure 4, the plotted line depicts the near-optimum bid recorded at each move during the entire simulation. This shows that lower costs of producing the components are gradually found as the currency values are adjusted iteratively. The first near-optimum bid was obtained at the 62nd move. The new job schedule for all machines is depicted in Table 4.

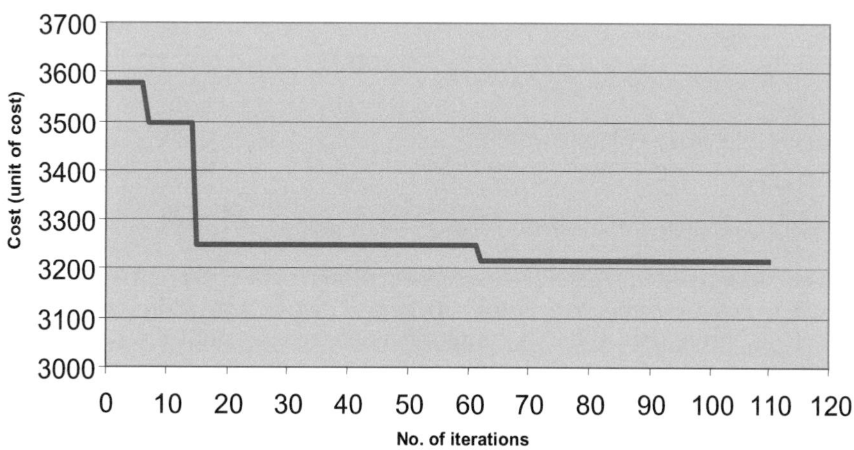

Fig. 4. Optimum bid recorded at each TS move ($\alpha = 15\%$).

Figures 5 and 6 show the results obtained in second test run, the bids received at each TS move and the optimum bid recorded during each move respectively. When the simulation completed, the near-optimum bid obtained was the same as the first test run and the first near-optimum bid was obtained at the 58th move. These results show that in many moves there are no bids received from the machine agents. This happens predominantly when diversification takes place. As each of the currency values is particularly allocated to a unique job to produce a particular feature, exchanging currency values from one job with another is inappropriate. For instance, assume that the currency values for the first feature of a component is relatively large (say 1500) and for the second feature is small (e.g. 500).

Lathe	Schedule	Milling	Schedule	Drilling	Schedule
L1	**ComA(2)* 212-470^ ComC(3) 530-788** *C4(2) 788-988*	M1	**ComB(1) 0-242 ComB(3) 446-515 ComA(3) 520-625** *C3(2) 625-917* **ComC(4) 945-1080** *C3(3) 1080-1276*	D1	**ComA(1) 0-180** *C1(1) 180-420* **ComC(5) 1200-1374**
L2	**ComC(1) 0-384 ComC(2) 384-522** *C1(2) 522-862*			D2	C2(2) 460-650
				D3	C2(4) 1200-1380
L3	C3(1) 0-415 C2(3) 750-925	M2	C2(1) 0-380 *C4(3) 1168-1336*		
L4	**ComB(2) 302-386 ComA(4) 728-844** *C1(4) 1186-1511*	M3	C4(1) 0-432 *C1(3) 912-1120*		

*C*x*(*y*) means job sequence *y* of component *x*
^ in unit of time
Highlighted in bold = new jobs being scheduled
Highlighted in Italic = existing jobs being rescheduled

Table 4. New machines job schedule

When diversification strategy (pairwise exchange) takes place, the currency value for the first feature is now 500 and as a result, there will not be any bids put forward by the machine agents throughout the entire intensification process until the next diversification takes place. However, the diversification strategy in TS leads to a great opportunity for the search process to explore new region aiming to obtain better solutions. This can be observed in Figure 3 during the moves from 90th to 105th that many bids have been put forward.

To evaluate further the effectiveness of the bidding mechanism, the simulation results obtained were further analysed by comparing with three heuristic integrated process planning and scheduling approaches by Khoshnevis & Chen (1993), Usher & Fernandes (1996) and Saygin & Kilic (1999). Khoshnevis & Chen (1993) proposed an integrated process planning and scheduling system whereby the two stages of process planning and production scheduling are treated as a unified whole. This system uses a six-step heuristic approach based on opportunistic planning to generate feasible process plans through the creation of detailed routing, scheduling and sequencing information. Usher & Fernandes (1996) proposed PARIS (Process planning ARchitecture for Integration with Scheduling) – a two-phased architecture for process planning that supports the integration with scheduling. Saygin & Kilic (1999) proposed a framework that integrates predefined flexible process plans with off-line (predictive) scheduling in flexible manufacturing systems.

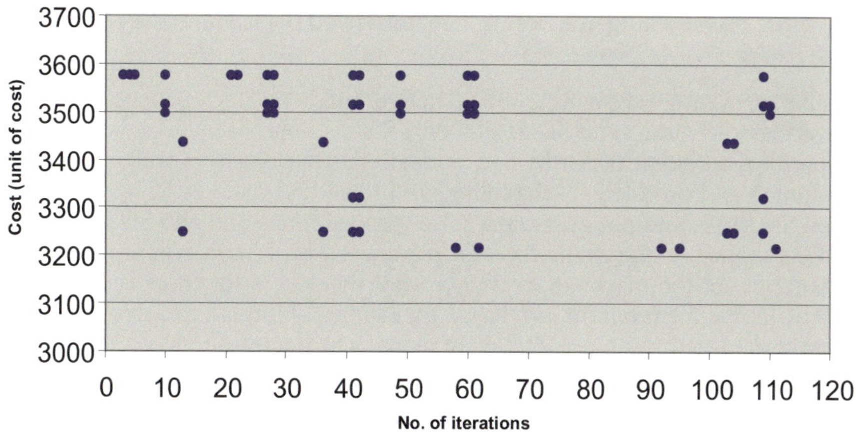

Fig. 5. Bids received at each TS move (α = 30%).

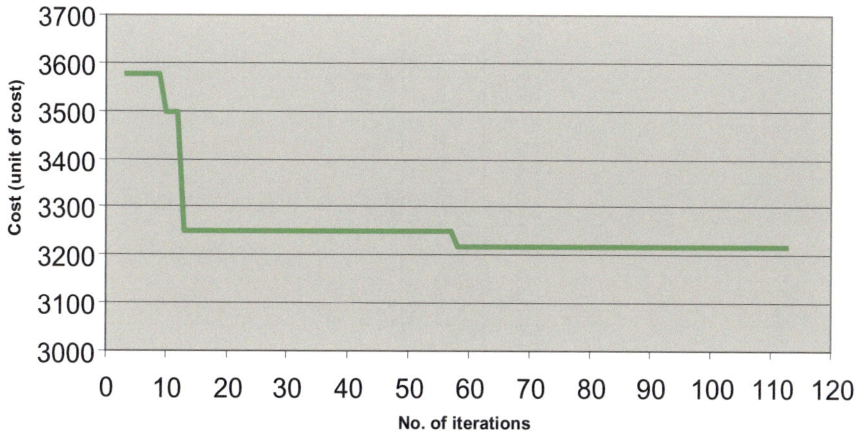

Fig. 6. Optimum bid recorded at each TS move (α = 30%).

In order to make a rational comparison with the iterative bidding MAS developed in this study, the same test case is used to simulate the three heuristics approaches. Based on the simulation results obtained, Table 5 can be drawn for comparison purposes between the four approaches. The highlighted sections indicate the best results between these approaches. Based on the results, the approach by Khoshnevis & Chen (1993) is not able to achieve more promising result (i.e., lower lead time and production cost) than the MAS. The results obtained for ComA and ComC are no better than those achieved by the MAS. However, this approach manages to achieve the same lead time and production cost for ComB as the MAS. For PARIS system, the static phase involves the determination of suitable processes for each feature and followed by machine-group selection, to produce a list of alternative process plans. In the dynamic phase, all of these alternative process plans are scheduled based on the operational status of the machine on the shop floor. In order to make a relevant comparison with the MAS, the criteria used in the process of scheduling are the production cost and lead time (i.e., to meet the delivery due dates). The results, once again, show that the MAS is able to obtain better results than this approach. Furthermore, this approach performs poorer than Khoshnevis & Chen (1993). As for Saygin & Kilic's approach, after rescheduling the results are improved which are the same as the ones obtained in Khoshnevis & Chen (1993).

These results noticeably show that the iterative bidding MAS proposed in this study outperforms these heuristic approaches. Not only it is capable of obtaining better solutions but also the way of the system performs (i.e., autonomous approach) is well suited to integrate process planning and production scheduling with time- and cost-efficiency. Unlike heuristics approaches, the MAS does not generate a list of process plans and allocate machines to these plans, and subsequently determine the best plan based on certain criteria. The MAS allows agents that represent the machines to decide what the best is for them (e.g. maximise their utilisation) by letting them bid for jobs based on their capability. In this way, a near-optimum solution can be achieved and the utilisation of manufacturing resources can also be optimised.

Component	**MAS**		**Khoshnevis & Chen**		**Usher & Fernandes**		**Saygin & Kilic (after rescheduling)**	
	Total Lead Time	Total Production Cost	Total Lead Time	Total Production Cost	Total Lead Time	Total Production Cost	Total Lead Time	Total Production Cost
ComA	**844**	**1894**	1254	2493	1276	2538	1254	2493
ComB	**515**	**1302**	**515**	**1302**	826	1475	**515**	**1302**
ComC	**1374**	**3224**	1882	3620	2430	4032	1882	3620

Table 5. Comparative results between MAS and heuristic approaches

7. Conclusion

In order to achieve manufacturing competence (through cost-competitiveness), this chapter introduced a multi-agent system (MAS) to enable process planning options and production scheduling options to be evaluated and optimised dynamically. The proposed MAS helps to enhance the agility and flexibility of manufacturing systems to cope with dynamic changes in the market by achieving near-optimum solutions to integrated process planning and scheduling problems. To achieve this, a novel currency-based iterative agent bidding mechanism is used as an agent coordination protocol. Agents representing the machines on the shop floor will bid for jobs to produce components; as iterative bidding takes place it aims to lead to better and better solutions to achieve cost-effectiveness.

To facilitate the iterative bidding mechanism, a Tabu search optimisation technique was developed to adjust the current values. A test case was used to simulate the agent bidding mechanism and test runs were executed to evaluate the effectiveness of the bidding mechanism. The simulation results show that as the currency values were adjusted at each TS move, the production cost of producing the components was gradually reduced. The results were then compared to the results obtained based on three heuristic approaches (Khoshnevis & Chen, 1993; Usher & Fernandes, 1996b; Saygin & Kilic, 1999). The comparative results show that the MAS outperforms the heuristic approaches. The MAS evaluates and optimises process plans and production schedules simultaneously. It allows agents that representing the machines to bid for jobs based on their capability and best performance (e.g. to maximise their machine utilisation). Furthermore, the MAS also provides a platform where the possible reconfiguration of manufacturing systems can be assessed and the utilisation of manufacturing resources can be optimised. For future work, the MAS could be enhanced with machine learning capability in order to facilitate the iterative bidding mechanism to achieve optimised solutions more rapidly and efficiently.

8. References

Blum, C. & Sampels, M. (2004). An Ant Colony Optimisation Algorithm for Shop Scheduling Problems. *Journal of Mathematical Modelling and Algorithm,* Vol.3, pp. 285–308.

Baykasoglu, A. & Gocken, T. (2010). Multi-Objective Aggregate Production Planning with Fuzzy Parameters. *Advances in Engineering Software,* Vol.41, No.9, pp. 1124-1131.

Baykasoglu, A. & Ozbakir, L. (2009). A Grammatical Optimization Approach for Integrated Process Planning and Scheduling. *Journal of Intelligent Manufacturing,* Vol.20, No.2, pp. 211-221.

Caridi, M. & Sianesi, A. (2000). Multi-Agent Systems in Production Planning and Control: An Application to the Scheduling of Mixed Model Assembly Lines. *International Journal of Production Economics,* Vol.68, pp. 29–42.

Chen, Q. & Khoshnevis, B. (1992). Scheduling with Flexible Process Plans. *Production planning and Control,* Vol.3, pp. 1-11.

Cho, S.Y. & Lazaro, A. (2010). Control theoretic model using PID controller for just-in-time production scheduling. *International Journal of Acvanced Manufactruing Technology,* Vol.51, Nos.5-8, pp. 699-709.

Demir, L.; Tunali, S. & Lokketangen, A. (2011). A Tabu Search Approach for Buffer Allocation in Production Lines with Unreliable Machines. *Engineering Optimisation,* Vol.43, No.2, pp 213-231.

Ferber, J. (1999). *Multi-Agent Systems: An Introduction to Distributed Artificial Intelligence,* England: Addison-Wesley.

Glover, F. (1989). Tabu Search: Part I. *ORSA Journal of Computing,* Vol.1, pp. 190-206.

Glover, F. (1990). Tabu Search: Part II. *ORSA Journal of Computing,* Vol.2, pp. 4-32.

Hajizadeh, Y.; Christie, M. & Demyanov, V. (2011). Ant Colony Optimization for History Matching and Uncertainty Quantification of Reservoir Models. *Journal of Petroleum Science and Engineering,* Vol.77, No.1, pp. 78-92.

Ho, Y.C. & Moodie, C.L. (1996). Solving Cell Formation Problems in a Manufacturing Environment with Flexible Processing and Routeing Capabilities. *International Journal of Production Research,* Vol.4, No.10, pp. 2901-2923.

Hou, T. & Wang, H. (1991). Integration of a CAPP System and an FMS. *Computers and Industrial Engineering,* Vol.20, No.2, pp. 231-242.

Hutchinson, G.K. & Pflughoeft K.A. (1994). Flexible Process Plans: Their Value in Flexible Automation Systems. *International Journal of Production Research,* Vol.32, No.3, pp. 707-719.

Iwata, K. & Fukuda, Y. (1989). A New Proposal of Dynamic Process Planning. *Proceedings of CIRP International Workshop on CAPP,* Hannover, Germany.

Khoshnevis, B. & Chen, Q. (1989). Integrated Process Planning and Scheduling Functions. *Proceedings of the IIE Integrated Systems Conference and Society for Integrated Manufacturing Conference,* pp. 415-420.

Khoshnevis, B. & Chen, Q. (1990). Integrated Process Planning and Scheduling Functions. *Journal of Intelligent Manufacturing,* Vol.1, pp. 165-176.

Khoshnevis, B. & Chen, Q. (1993). Scheduling with Flexible Process Plan. *Production Planning and Control,* Vol.4, No.4, pp. 333-343.

Kim, H.J.; Chiotellis, S. & Seliger, G. (2009). Dynamic Process Planning Control of Hybrid Disassembly Systems. *International Journal of Advanced Manufacturing Technology,* Vol.40, Nos.9-10, pp. 1016-1023.

Kiritsis, D. & Porchet, M. (1996). A Generic Petri Net Model for Dynamic Process Planning and Sequence Optimization. *Advances in Engineering Software,* Vol.25, pp. 61-71.

Kumar, N.; Tiwari, M.K. & Chan, F.T.S. (2008). Development of a Hybrid Negotiation Scheme for Multi-Agent Manufacturing Systems. *International Journal of Production Research,* Vol.46, No.3, pp. 539–569.

Larsen, N.E. & Alting, L. (1990). Simultaneous Engineering within Process and Production Planning. *Pacific Conference on Manufacturing, Sydney and Melbourne,* pp. 1024-1031.

Li, X.Y.; Shao, X.Y.; Gao, L. & Qian W.R. (2010). An Agent-Based Approach for Integrated Process Planning and Scheduling. *Expert Systems with Applications,* Vol.37, No.2, pp. 1256-1264.

Lima, R.M.; Sousa, R.M. & Martins, P.J. (2006). Distributed Production Planning and Control Agent-Based System. *International Journal of Production Research,* Vol.44, Nos.18/19, pp. 3693–3709.

Liu, Q.; Lv, L. & Feng, T.T. (2007). Research on Reconfiguration of Manufacturing Resource Based on Multi-Agent. *Journal of Wuhan University of Technology,* Vol.29, No.12, pp. 119–122.

Mallur, S.; Mei, J. & Zhang, H. (1992). An Introduction to an Integrated Process Planning. *ASME Manufacturing International*, pp. 335-347.

Mamalis, A.G.; Malagardis, I. & Kambouris, K. (1996). On-Line Integration of a Process Planning Module with Production Scheduling. *International Journal of Advanced Manufacturing Technology*, Vol.12, No.5, pp. 330-338.

Moon, C.; Lee, Y.H.; Jeong, C.S. & Yun, Y.S. (2009). Integrated Process Planning and Scheduling in a Supply Chain. *Computer and Industrial Engineering*, Vol.57, No.4, pp. 1484-1484.

Moslehi, G.; Mirzaee, M.; Vasei, M.; Modarres, M. & Azaron, A. (2009). Two-Machine Flow Shop Scheduling to Minimize the Sum of Maximum Earliness and Tardiness, *Interntional Journal of Production Economics*, Vol.122, No.2, pp. 763-773.

Omar, M.K. & Teo, S.C. (2007). Hierarchical Production Planning and Scheduling in a Multi-Product, Batch Process Environment. *International Journal of Production Research*, Vol.45, No.5, pp. 1029-1047.

Ottaway, T.A. & Burns, J.R. (2000). An Adaptive Production Control System Utilising Agent Technology. *International Journal of Production Research*, Vol.38, No.4, pp. 721–737.

Robu, V.; Noot, H.; La Poutre, H. & van Schijndel, W. (2011). A Multi-Agent Platform for Auction-Based Allocation of Loads in Transportation Logistics. *Expert Systems with Applications*, Vol.38, No.4, pp. 3483-3491.

Saygin, C. & Kilic, S.E. (1999). Integrating Flexible Process Plans with Scheduling in Flexible Manufacturing Systems. *International Journal of Advanced Manufacturing Technology*, Vol. 15, No. 4, pp. 268-280.

Srihari, K. & Greene, T.J. (1988). Alternative Routeings in CAPP Implementation in a FMS. *Computers and Industrial Engineering*, Vol.15, Nos.1-4, pp. 41-50.

Ssemakula, M.E. & Wesley J.C. (1994). Functional Specification of a Dynamic Process Planning System. *Computers and Industrial Engineering*, Vol.27, Nos.1-4, pp. 99-102.

Tonshoff, H.K.; Beckendorff, U. & Anders, N. (1989). FLEXPLAN: A Concept for Intelligent Process Planning and Scheduling. *Proceedings of CIRP International Workshop on CAPP*, pp. 87-106.

Usher, J.M. & Fernandes, K. (1996). A Two-Phased Approach to Dynamic Process Planning. *Computers and Industrial Engineering*, Vol.31, Nos.1/2, pp. 173-176.

Wang, J.; Zhang, Y.F. & Nee, A.Y.C. (2009). Reducing Tardy Jobs by Integrating Process Planning and Scheduling Functions, *International Journal of Production Research*, Vol.47, No.21, pp. 6069-6084.

Wang, L.H.; Feng, H.Y. & Cai, N.X. (2003). Architecture Design for Distributed Process Planning. *Journal of Manufacturing Systems*, Vol.22, No.2, pp. 99-115.

Wang, X.; Li, Z., Liu, C. & Tian, W. (1995). Process Planning Under Job Shop Environment. *SPIE*, 2620, pp. 143-148.

Wang, Y.F.; Zhang, Y. & Fuh J. Y. H. (2011). Job Rescheduling by Exploring the Solution Space of Process Planning for Machine Breakdown/Arrival Problems. *Proceedings of the Institution of Mechanical Engineers, Part B: Journal of Engineering Manufacture*, Vol.225, No.2, pp. 282-296.

Wong, T.N.; Leung, et al. (2006). Dynamic Shop Floor Scheduling in Multi-Agent Manufacturing Systems. *Expert Systems with Applications*, Vol.31, pp. 486–494.

Wooldridge, M. (1997). Agent-Based Software Engineering. *IEE Proceedings on Software Engineering*, Vol.144, No.1, pp. 26-37.

Xu, K.; Feng, Z. & Jun, K. (2010). A Tabu-Search Algorithm for Scheduling Jobs with Controllable Processing Times on a Single Machine to Meet Due-Dates. *Computers and Operations Research*, Vol.37, No.11, pp. 1924-1938.

Zhang, H. (1993). IPPM-A Prototype to Integrate Process Planning and Job Shop Scheduling Functions, *CIRP*, Vol.42, No.1, pp. 513-518.

Zhang, Z. (2011). Towards Theory Building in Agile Manufactruing Strategies - Case Studies of an Agility Taxonomy. *International Journal of Production Economics*, Vol.131, No.1, pp. 303-312.

10

Using Timed Coloured Petri Nets for Modelling, Simulation and Scheduling of Production Systems

Andrzej Bożek
Rzeszow University of Technology,
Poland

1. Introduction

The right choice of a system model is the important issue related to developing of production management systems. An abstract representation of the system is needed for e.g. production scheduling or "what-if" scenarios generating. Disjunctive graphs, permutations with repetition as well as mathematical programming relations are the most common used representations. However, these representations make possible to model systems with rather simple and homogenous structures. Different extensions are added to models, as setup and transport times, machine calendars, resource restrictions, etc., to represent other system features. These models are often carefully designed and find application in robust scheduling algorithms. Sometimes, however, the high efficiency of the scheduling is not the primary goal, but a flexible modelling formalism is more important. This chapter proposes a solution dedicated especially for these cases.

Petri nets are successfully used for modelling and scheduling of production systems. The presented solution is based on hierarchical timed coloured Petri nets (HTCPN), the formalism introduced by Jensen. This formalism has been chosen because of its generality and flexibility. Token colouring lets tokens carry information that can be represented by complex data types and can be processed by arc expressions. The hierarchy allows to decompose a structure of a system to nested modules. Multiple instances of an once developed module can be parametrised and placed into a model many times. Timestamps assigned to tokens and the model global clock make possible to simulate the time flow that is especially important for scheduling.

2. Petri nets in production scheduling

Petri nets are often used approach in modelling, simulation, analyzing and scheduling of production systems. The excellent review has been prepared by Tuncel and Bayhan (2007). The authors verified 72 publications that concern using Petri nets in the production scheduling. The works have been differed on the basis of a scheduling approach, a Petri net type and an application area. Four main scheduling approaches have been enumerated: heuristic rule based systems, search algorithms, mathematical programming approaches and meta-heuristics. Very different variants of Petri net formalism are used, there are *stochastic Petri nets* (SPN), *priority nets, coloured transition-timed Petri nets, object-oriented Petri nets* (OOPN), *generalised symmetric/asymmetric nets* (GSN/GAN), and others. The most authors use Petri nets for modelling and scheduling of *flexible manufacturing systems* (FMS). However, the definition of the FMS is very general and the detailed application can be unknown. The other areas of application are, for example, job shop scheduling, AGV systems, cyclic scheduling, wafer fabrication, demanufacturing systems. In the breakdown of the 72 publications, at least 13 of them refer to timed Petri nets, and at least 7 refer to coloured Petri nets.

The work published by Camurri, Franchi and Gandolfo (1991) is probably one of the first in which both coloured and timed Petri nets have been used for production modelling and scheduling. The authors use coloured tokens to distinguish scheduled jobs. It is used a Petri net variant that assigns enabling times to transitions. These enabling times model processing times of operations. There are also the *executor subsystem* for a net model simulation and the *scheduling subsystem* for scheduling decision making proposed in the work.

Aized (2010) refers to *Coloured Petri Net* (CPN) formalism introduced by Jensen (Jensen & Kristensen, 2009). The author emphasizes the flexibility of CPN in modelling and scheduling of *Discrete Event Dynamical Systems* (DEDS). He discusses the model of a semi-conductor manufacturing system (called *multiple cluster tool system*) that has been implemented with the use of CPN formalism and *CPN Tools* software.

In the work of Zhang, Gu and Song (2008), the Jensen's CPN formalism has been also used for modelling of production systems. The authors place elements that represent batch processing, setup operations and transport operations in a one net structure.

3. Formalism of hierarchical timed coloured Petri nets

There are many variants of the Petri net formalism. Basic Petri nets, so-called *low-level nets* or *P/T nets*, have the syntax based on places, transitions and arcs. In practice, *high-level Petri net* formalisms are often used. These formalisms include additional extensions, e.g. colouring, time representation, special types of arcs, priority of transitions, hierarchy, and others. In this work, the formalism of *Coloured Petri Nets* introduced by Jensen is used (Jensen & Kristensen, 2009). There are two main reasons of that choice:

1. This formalism has an extensive syntax and semantics that makes modelling very comfortable and flexible.
2. There is a software called *CPN Tools* (CPN Tools Homepage, 2011) that fully supports the *Coloured Petri Nets* formalism. *CPN Tools* makes possible to edit, simulate and analyse of Petri net models.

Colouring, time representation and hierarchy representation are the main features of the formalism of *Coloured Petri Nets*. Because of that, this formalism will be denoted shortly as HTCPN (*Hierarchical Timed Coloured Petri Net*) in the remainder of this chapter.

The formal definition of the syntax of HTCPN, except for hierarchy representation, is as follows (Jensen & Kristensen, 2009):

A timed non-hierarchical coloured Petri net is a nine-tuple $CPN_T = (P, T, A, \Sigma, V, C, G, E, I)$ where:

1. P is a finite set of *places*.
2. T is a finite set of *transitions* such that $P \cap T = \emptyset$.
3. $A \subseteq P \times T \cup T \times P$ is a set of direct *arcs*.
4. Σ is a finite set of non-empty *colour sets* (types). Each colour set is either untimed or timed.
5. V is a finite set of *typed variables* such that $Type[v] \in \Sigma$ for all variables $v \in V$.
6. $C : P \to \Sigma$ is a *colour set function* that assigns a colour set to each place. A place p is timed if $C(p)$ is timed, otherwise p is untimed.
7. $G : T \to EXPR_V$ is a *guard function* that assigns a guard to each transition t such that $Type[G(t)] = Bool$.
8. $E : A \to EXPR_V$ is an *arc expression function* that assigns an arc expression to each arc a such that
   - $Type[E(a)] = C(p)_{MS}$ if p is untimed (*MS*: multi set),
   - $Type[E(a)] = C(p)_{TMS}$ if p is timed (*TMS*: timed multi set),
   where p is the place connected to the arc a.
9. $I : P \to EXPR_{\emptyset}$ is an *initialisation function* that assigns an initialisation expression to each place p such that
   - $Type[I(p)] = C(p)_{MS}$ if p is untimed,
   - $Type[I(p)] = C(p)_{TMS}$ if p is timed.

The hierarchy feature has been omitted in the definition above, because its strict formulation is quite complex. However, this concept is intuitively very simple. Selected substructures of a net model can be placed in special blocks called *modules*. Any module can be inserted to the net structure many times, where it is represented by a special kind of transition, so-called *substitution transition*.

The formal definition of the HTCPN semantics is also quite complex. It is mainly based on concept of timed multisets. This definition will not be quoted here. Instead, a few simple examples of HTCPN structures will be presented that demonstrates basics of the semantics. The first of the structures (fig. 1a) represents a model equivalent to a low-level net. From the formal point of view, it is a coloured net because place markings and arc expressions are typed. However, the type (colour set) of all places and arcs is *UNIT*. It is the simplest type predefined in the *CPN Tools* environment that has only one value, denoted by the empty pair of brackets (). A token of type *UNIT* can be either present or absent and it does not carry any additional information. Thereby, *UNIT* type tokens emulate indistinguishable black tokens and the net structure presented in the figure 1a is semantically equivalent to a basic P/T net. An expression that represents n tokens of type *UNIT* has the form *n*\`(), if n = 1, it can be simplified to the notation (). The discussed net (fig. 1a) has 7 places and 4 transitions. The initial marking assigns 1 token to the places *free_A* and *free_B* as well as 4 tokens to the place *jobs*. Every arc carries 1 token at a time. The structure can be interpreted as a model of a simply production system in which 2 machines (*A* and *B*) have to process 4 jobs. Every machine has 2 states (*free* and *ready*). Changes between the states are modelled by moving tokens when transitions *setup* and *process* are executed. These executions represent events in the system. It is obvious that only a one transition from the pair (*setup_A*, *process_A*) and also (*setup_B*, *process_B*) can be enabled (that means ready for execution) at a time. Thus, at the most two transitions can be enabled simultaneously in the whole net. If two or more transitions are enabled, the selection of that one to be executed is nondeterministic. This rule is an important characteristic of Petri nets that makes possible to determine many variants of evolution of a concurrent discrete event system. Different variants can be a result of a lack of full information about a system or it can be an immanent feature of a modelled system. In the result of the execution of the model presented in the figure 1a, it is possible to get different final marking in the places *result_A* and *result_B* with the restriction that the total number of tokens in these places has to be 4. This relates to the fact that times of processes and setups are not known, so it is undetermined how many jobs will be processed by any of machines. The final marking denoted in the figure 1a (rectangles close to places) is only one of the possibilities.

The model in the figure 1b represents a timed Petri net. This net structure is similar to that previously discussed (fig. 1a) but a timed extension has been added. The time representation is based on two elements in the HTCPN formalism. Token *timestamps* are the first element and the *global clock* is the second one. The HTCPN net can have either untimed and timed places. Every token in a timed place has attached a special value called *timestamp*. The condition of transition enabling is most restricted for the timed HTCPN than for untimed. A transition is consider to be enabled if and only if there are sets of tokens in its input places sufficient to cover demand of all input arcs (identically as for an untimed net), but for timed places only the tokens having timestamps not greater than a value of the global clock are taken for this calculation. In other words, a timed token can be taken from a place only if its timestamp is less than or equal to a value of the global clock.

Respecting given enabling definition, the algorithm of a timednet execution can be formulated as follows:

1. Set the global clock to 0.
2. If there are enabled transitions go to 4 else go to 3.
3. Increase the value of the global clock to the smallest possible value for which there are enabled transitions, if it is impossible then STOP.
4. Select randomly a subset of the enabled transitions (more precisely it should be a subset of so-called *binding elements*).
5. Execute selected transitions.
6. Go to 2.

The notation <*untimed marking*>@*TS*, used for defining a marking of timed places, denotes that tokens represented by the expression <*untimed marking*> have the timestamp *TS*. It is thought $TS = 0$ if @*TS* postfix is omitted. The notation <*untimed marking*>@+*d* on an input arc of a timed place denotes that tokens represented by the expression <*untimed marking*> will be inserted to the destination place with the timestamp $TS = GC + d$, where *GC* is the value of the global clock in the moment when the transition connected with the arc is executed. The inscription @+*d* can be also added to a transition and then the value *d* applies to all output arcs of the transition. The parameter *d* is called *delay*. It can be informally thought as a time of the event modelled by the execution of the transition.

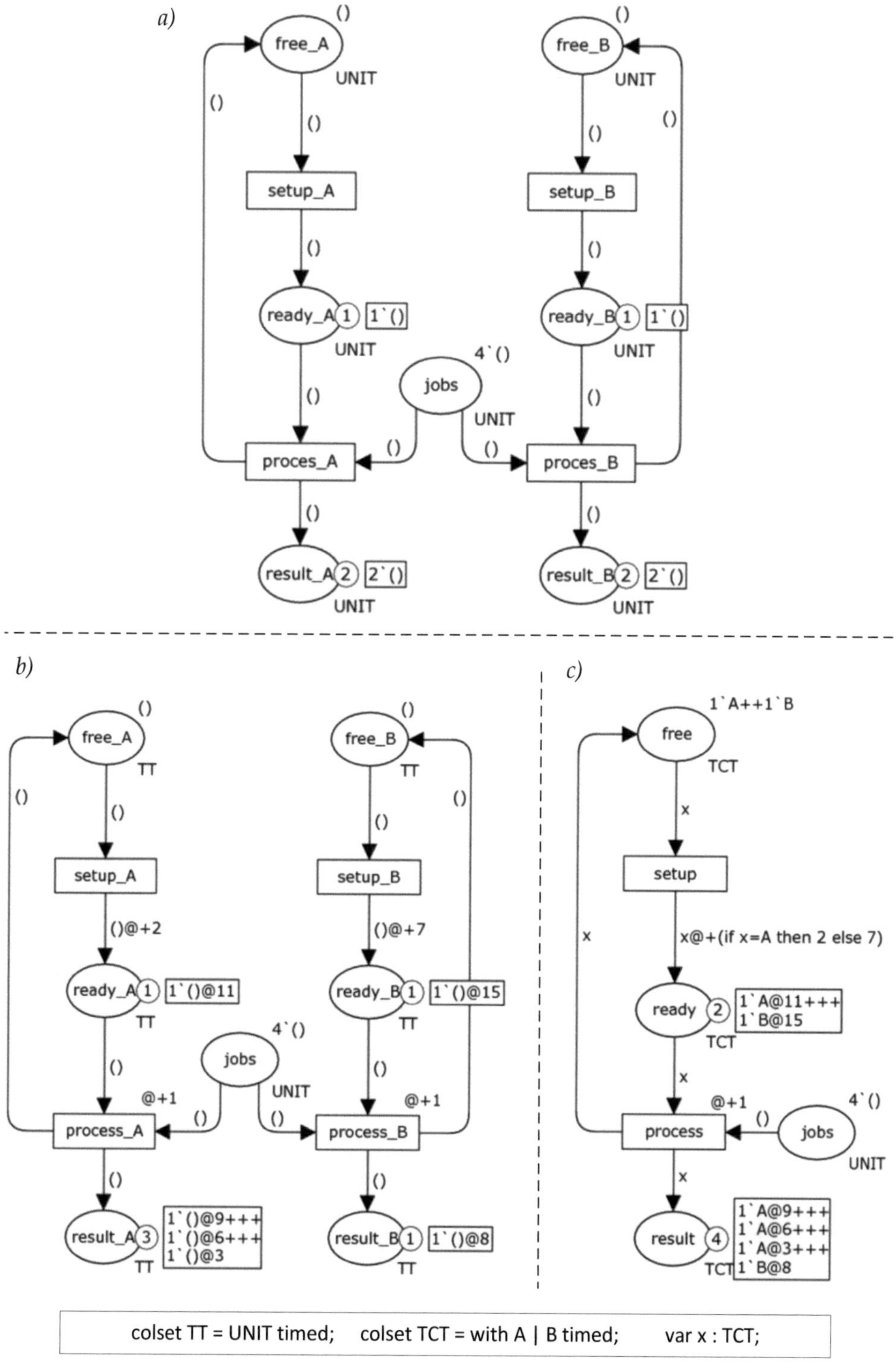

Fig. 1. The example CPN structures: a) simple P/T net, b) timed net, c) timed coloured net.

In the model presented in the figure 1b, the timed type *TT* (*Timed Type*) is assigned to all places, except *jobs*. This type represents black tokens (*UNIT*) with timestamps. The initial marking is the same as in the previous model (fig. 1a), but the tokens in the places *free_A* and *free_B* get implicitly timestamps equal 0. Arc and transition delays define duration of modelled activities: the setup A lasts 2 (units of time), the setup B lasts 7 and the both processes last 1. At the start of the model execution, two enabled transitions *setup_A* and *setup_B* are fired (with the nondeterministic selection). After that, the place *ready_A* has the marking 1`()@2 and *ready_B* has the marking 1`()@7. Next, the global clock is increased to the value 2 and the transition *process_A* becomes enabled and it is executed. After this execution, the place *free_A* has the marking 1`()@3 and the marking of the place *free_B* remains 1`()@7, thus, the global clock is increased to the value 3 and the transition *setup_A* is executed, and so on. The final marking of the net model (fig. 1b) specifies that the machine *A* will process three jobs, that will finish in the moments 3, 6 and 9, while the machine *B* will process a one ajob to the moment 8. An analysis or a detailed simulation of the net model shows that it is only one possible final marking in this example.

Colouring of the HTCPN is syntactically represented by assigning colour sets (types) to places, defining typed variables and assigning expressions to arcs and transitions. It introduces elements of a textual programming language with complex data types to the Petri nets modelling. The syntax definition does not specify the language exactly. In *CPN Tools* the functional programming language called *CPN ML* is used, that is an extension of the *Standard ML* language. There are two main advantages of using colouring in Petri nets:

1. It is possible to build more compact structures, because similar substructures can be folded to a one subnet by replacing structural diversity with the diversity of token values.
2. It is possible to execute procedures of arithmetic computation and data processing that would be complex or even unfeasible with the use only structural net elements.

The example of a coloured HTCPN model has been shown in the figure 1c. The model is fully functionally equivalent to the one previously discussed (fig. 1b). Each pair of places (*free_A, free_B*), (*ready_A, ready_B*), (*result_A, result_B*) has been replaced with a one place, respectively: *free, ready* and *result*. The modelled machines are distinguished by token values. Each token has one of the two values *A* or *B* with a timestamp, that is specified by enumerated colour set definition *TCT* (*Timed Coloured Type*). Variables assigned to input arcs of transitions are often used in the coloured net models. These variables bind a value of a token consumed by a transition and this value can be used in output arc expressions. For example, the expression *x+@(if x = A then* 2 *else* 7) has been used in the considered model to infer a delay value from a machine type.

4. Modelling production systems by hierarchical timed coloured Petri nets

The HTCPN is a powerful formalism for modelling and simulation of production systems. In this heading the HTCPN patterns and substructures will be introduced that reflect some features of production systems (sequence constraints, batch processing, buffer limits, etc.). These patterns and substructures are in majority loose coupled, so can be added to models independently. Some of the concepts presented in the next subheadings have been adopted from a literature and others have been fully introduced by the author. All the concepts form a compact review that presents how to construct a HTCPN model of even complex production system.

4.1 Production system structure

There are used many different models of production systems. The popular models considered in the context of modelling and scheduling problems are *single machine models, parallel machine models, flow shop, job shop* and *open shop* (Pinedo, 2008). It is possible to find applications of Petri nets in problems connected with various models. However, in this chapter, the model called *flexible job shop* (FJS) (Pinedo, 2008) has been taken into consideration. There are two reasons of this choice:

1. FJS is the strongly generalized model of a production system. It covers flow shop, job shop and parallel machine environments. So, the FJS can be used for modelling many of systems, including majority of *flexible manufacturing systems* (Christo & Cardeira, 2007).
2. The presented modelling concepts have been put into practice in a screw factory, where the production system has the FJS structure.

Models of production systems with sequence constraints, also the FJS model, are often represented by the disjunctive graph formalism (Brucker, 2007). A disjunctive graph of a hypothetical FJS structure is presented in the figure 2a. There are 3 machines (X, Y, Z) and 3 jobs (A, B, C) in the system. The job A has 2 operations, the first of them is processed on the machine X (and it lasts 45), the second one can be processed on the machine Y (30) or Z (25). Similarly, the job B is processed on the machine X (30) or Y (110) and next on the machine Z (40), whereas the job C is processed on machines X (50), Y (45) and Z (15) in turn.

It is easy to construct the HTCPN model that represents the same FJS system as a given disjunctive graph. For example, the graph in the figure 2a and the net model in the figure 2b relate to the same FJS system. The rules of the net model construction has been assumed as follows:

1. For each operation there is a place *init_<operation>*.
2. For each job there is a place *end_<job>*.
3. For each machine there is a place *mach_<machine>*.
4. For each pair operation-machine there is a transition *proc_<operationMachine>*. The transition has a time delay inscription @+d, where d is the processing time for the pair operation-machine.
5. For each transition *proc_<operationMachine>* there are:
   - The arc directed from the place *init_<operation>* to the transition,
   - The arc directed from the transition to the place *init_<nextOperation>*, if there exists a next operation in a job, or else to the place *end_<job>*.
   - The bidirectional arc that connects the transition and the place *mach_<machine>*.
6. All the places have the colour set *JOB*, that is defined as *UNIT timed.*
7. All the places *init_<operation>* that represent the first operations in jobs and also all the places *mach_<machine>* have initial marking (), all other places have empty initial marking.
8. All the arcs have the arc expression ().

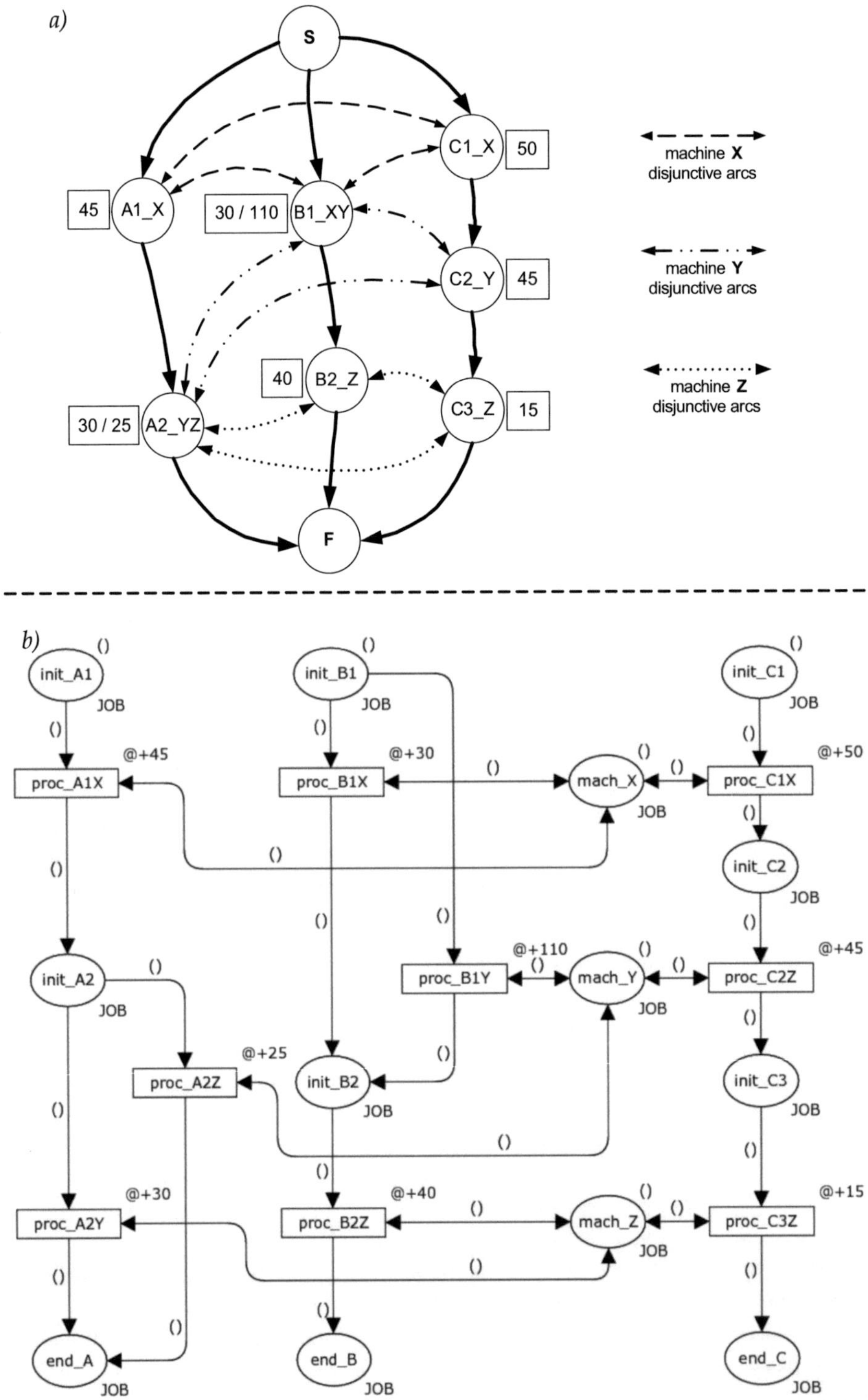

Fig. 2. The representation of the FJS system: a) disjunctive graph, b) timed Petri net model.

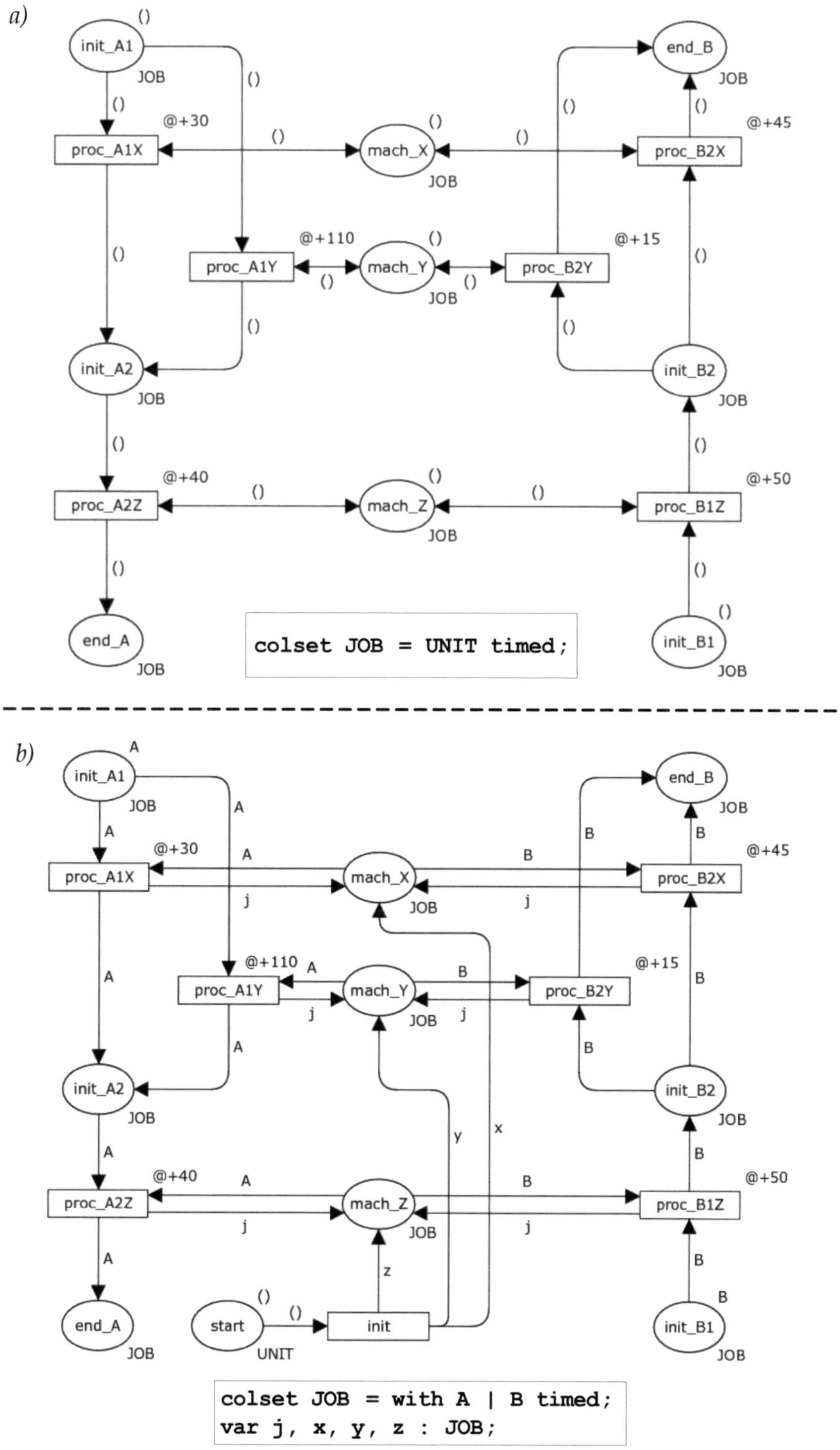

Fig. 3. Different sequencing rules: a) non-delay sequences, b) all sequences.

For each job tokens flow from the *init* place of the first operation, through consecutive operation places, to the *end* place. The delay added to timestamps as a result of executing of *proc* transitions causes that no operation can start until a previous one finishes and a machine token is blocked while operation processing.

It can be noted that the FJS system representations in the figures 2a and 2b are not completely functionally equivalent. While the disjunctive graph (fig. 2a) represents all possible operation sequences, the net structure (fig. 2b) models only so-called *non-delay* schedules (Pinedo, 2008). It is because the *mach* places are marked by non-coloured tokens and operations to processing cannot be reserved in advance. So, if any operation is waiting for a machine and a machine is free, processing starts immediately. However, it is also not difficult to build the net structure that models all sequences and is fully equivalent to disjunctive graph. An example is shown in the figure 3b. The models in the figures 3a and 3b represent the same simple production system, but the first one has been constructed on the basis of the previously given rules and the second one has been prepared to generate all sequences. The type *JOB* has been redefined (fig. 3b) to have a set of values (in the example *A* and *B*) that represent processed jobs. Tokens coming in to the *mach* places have randomly selected colours (*A* or *B*), because these tokens are generated by so-called *free variables* (Jensen & Kristensen, 2009). So, a sequence of operations for each machine is selected non-deterministically, and thus any execution scenarios can be achieved.

4.2 Batch processing

It is not very challenging and useful to create HTCPN models equivalent to disjunctive graphs, because there are large number of effective simulation and scheduling algorithms that use the disjunctive graph representation. The usefulness of modelling by Petri nets relates mainly to flexibility in adding different extensions to a model. Some examples of that extensions are presented in the remainder of this heading.

Operations do not have to consist of a single processing action. An initial marking of the first *init* place of a job can have many tokens instead of one. Moreover, arcs connected to *init* places can also carry many tokens at a time. This extension makes possible to model batch processing. In any operation some amount (batch) of a job is processed. That processing amount can be assign individually for each machine-job pair. An example is presented in the figure 4a. The job *A* has the total amount 50. In the first operation of the job *A* the amount 5 is processed, if it is done on the machine *X*, or the amount 15, if it is done on the machine *Y*. In the second operation of the job *A* the amount 10 is processed, and so on. The concept of modelling of batch processing presented in the figure 4a is adopted from the solution proposed by Zhang, Gu and Song (2008).

It is a possible situation that not all amount of a job will be processed, because a last batch in the *init* place will be smaller than a processing amount of every machine. That is unavoidable for most combinations of real-world parameters. It does not have to be a problem. If an amount is large and it is processed in many batches, a small rest at the end of the processing could be probably neglected. But if it is necessary to process full amount, the solution shown in the figure 4b can be used. The additional place (*ctr_A1*) has been introduced that is connected to all *proc* transitions related to a one operation (here *A1*) and this place counts the amount that remains unprocessed. If this amount is smaller than a standard batch size for a machine the smaller last batch is processed.

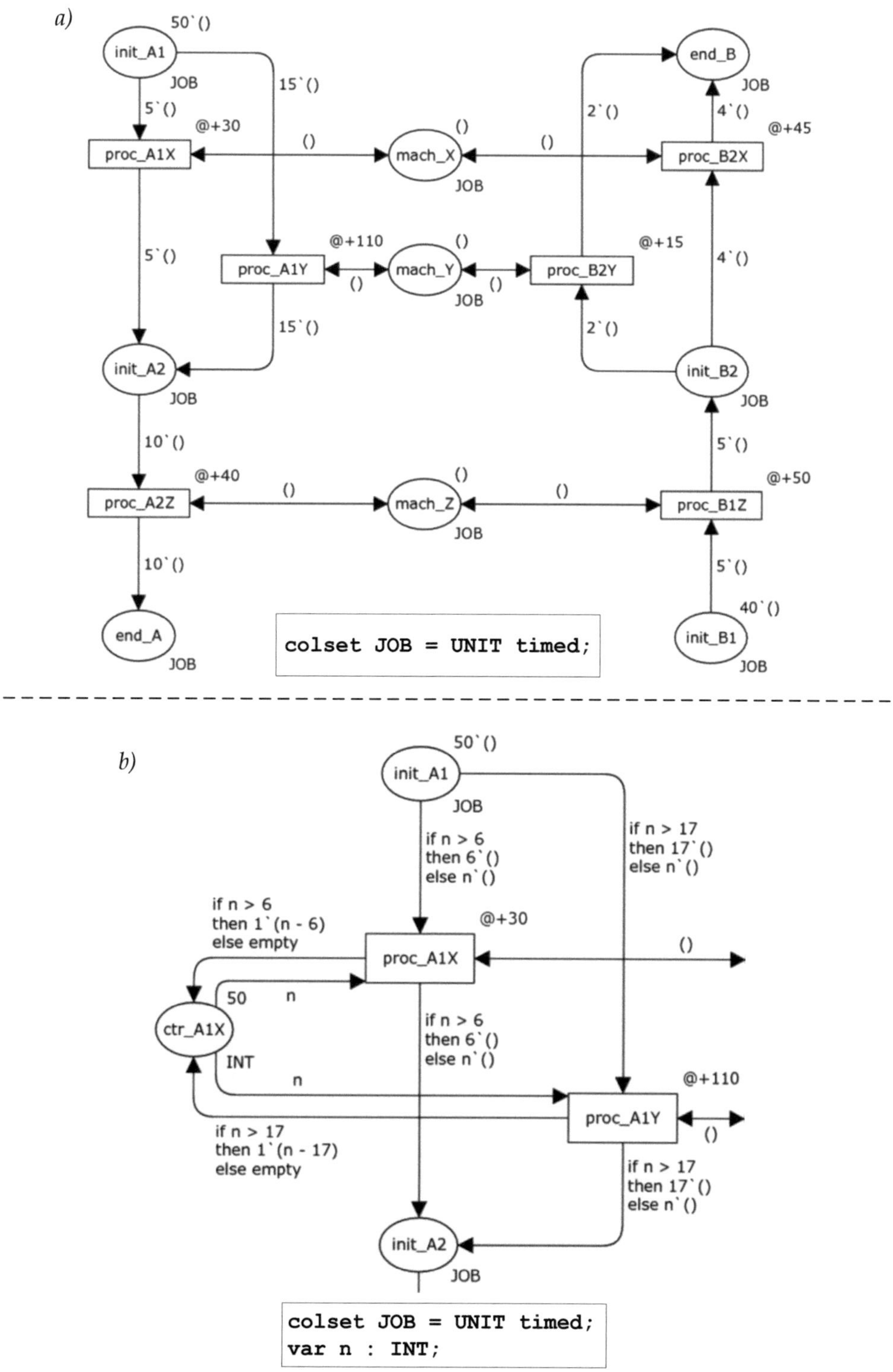

Fig. 4. Modelling batch production: a) Petri net model, b) processing of smaller last batch.

4.3 Extended machine model

The simple machine representation proposed in the prior subheadings can be insufficient for modelling of real production. Additional operation rules and parameters have to be allowed for. For example, an extended machine model is presented that combines following features:

1. Setup times between operations are taken into consideration.
2. An availability calendar is defined for a machine.
3. A machine processes operations batch by batch. Batches are processed without interrupting, so a batch processing starts only if it can be finished in the same bracket of availability.

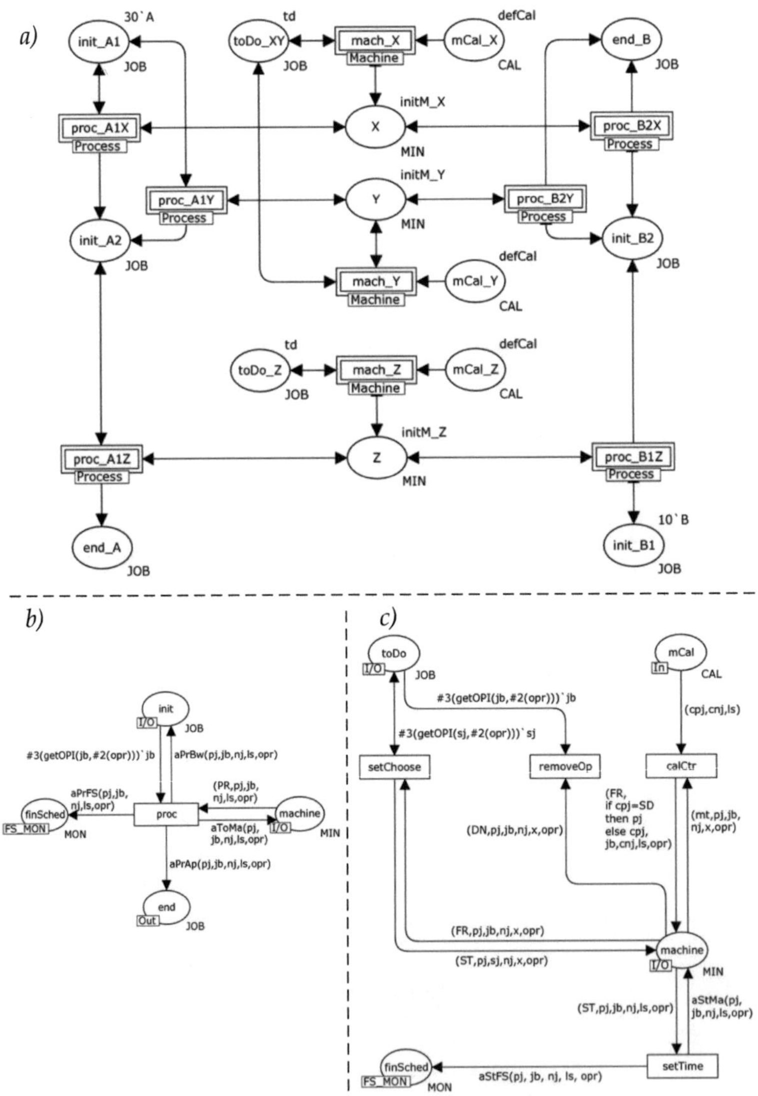

Fig. 5. The HTCPN structure with an extended machine model: a) the top-level structure, b) the inside view of *Process* module, c) the inside view of *Machine* module.

The HTCPN model of a simple production system with the extended machine model is shown in the figure 5. The hierarchical two-level net structure has been used. Details of substructures that represent batch processing logic and machine model logic has been moved to the modules *Process* (fig. 5b) and *Machine* (fig. 5c). Thus, the top-level structure (fig. 5a) remains quite simple and legible.

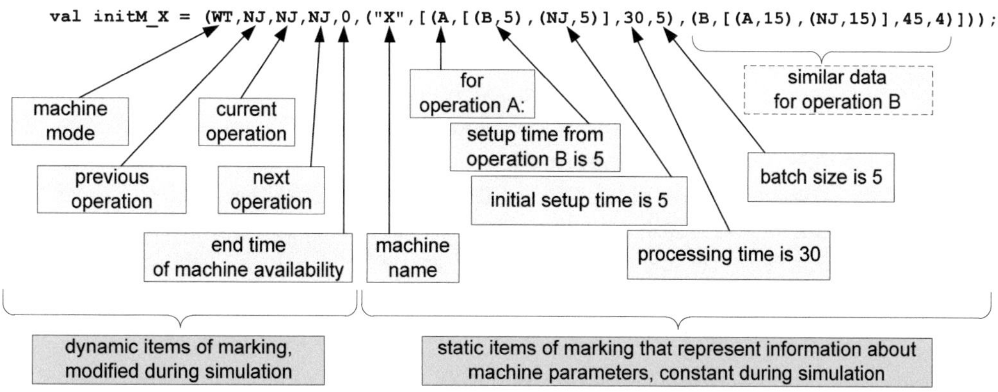

Fig. 6. The example definition of a token that represent machine state and parameters.

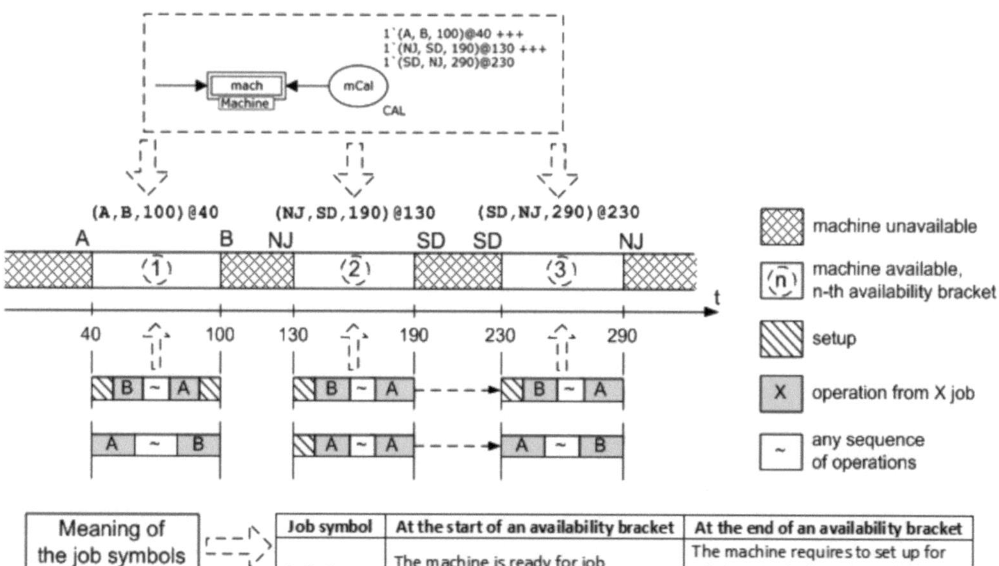

Job symbol	At the start of an availability bracket	At the end of an availability bracket
A, B, C, ...	The machine is ready for job A, B, C ... at the start of a bracket	The machine requires to set up for job A, B, C ... before the end of a bracket
SD (*Stopped*)	The machine was stopped during a finishing unavailability bracket so it is set up for the last job of a previous availability bracket	The machine will be stopped during a coming unavailability bracket
NJ (*None Job*)	The machine is not ready for any job	The machine does not require to set up for any job before the end of a bracket

Fig. 7. The representation of a machine availability calendar by a substructure of a HTCPN model.

Each place that represents a machine state (*X*, *Y*, *Z* in fig. 5a) is connected to the *Machine* substitution transition. These transitions include internal logic of machines. Additional places *toDo* control amount of batches that remains to be processed. Places *mCal* (a one place per a machine) represent calendars of machines availability. The structures of the modules *Process* and *Machine* are quite complex and these will not be detailed discussed here.

The marking of the places *X*, *Y*, *Z* consists of two sets of items (fig. 6). The first set groups items that represent dynamic elements of a machine logic. These items specify a present state of a machine, executed operations (previous, current and next) and end time of a machine availability bracket. This information is needed for control a simulation of the machine work. The second set groups items that represent parameters of machines that remain constant during the simulation. These parameters specify setup times for all possible previous operations, a processing time (per batch) and a batch size for each operation that the machine can execute.

The illustration in the figure 7 explains the HTCPN implementation of a machine availability calendar. The initial marking of the *mCal* place defines the set of availability brackets. The brackets are defined not only by start and end moments, but also by start and end job symbols. These symbols are needed to determine setup operations at the starts and at the ends of the brackets. The presented calendar substructure can be easily modified or extended to represent varying efficiency of a machine, periodical changes of availability, etc.

4.4 Transport subsystems

It is often necessary to model also transport processes in a production system. Transport is usually an activity realized between two consecutive operations. So, this activity can be modelled by additional net elements inserted between transitions that represent operations before and after transport (fig. 8a). In the simplest case, it can be a one transition with time delay (fig. 8a, *trans_A12*). This transition adds a transport delay to carried tokens. It should be noted that the transition is fired immediately when a sufficient number of tokens waits in an input place. Thus, intervals of transport time of many batches can overlap (fig. 8a). This is the behaviour that characterize for example a conveyor belt. Of course, the size of process batch and the size of related transport batch don not have to be equal.

If dedicated transport is used, for example a forklift truck, it can be modelled as shown in the figure 8b. The transition *trans_A12* carries tokens from the place *fin_A1* to the place *init_A2* and simultaneously it adds delay to the timestamp of the token in the place *TRANS*. Because of that, no following transport can start until the delay time expires. If this delay time is greater than transport time, the difference represents return time of the vehicle. For example, the transport modelled in the figure 8b lasts 50 and return time is 30.

In the figure 8c, a hypothetical situation is presented where three operations share a pair of transport vehicles (two tokens in the place *TRANS*). Transport between the operations *A1* and *A2*, as well as *B2* and *B3* can be realized by any one of the vehicles. However, the transport between the operations *A2* and *A3* requires two vehicles at the same time.

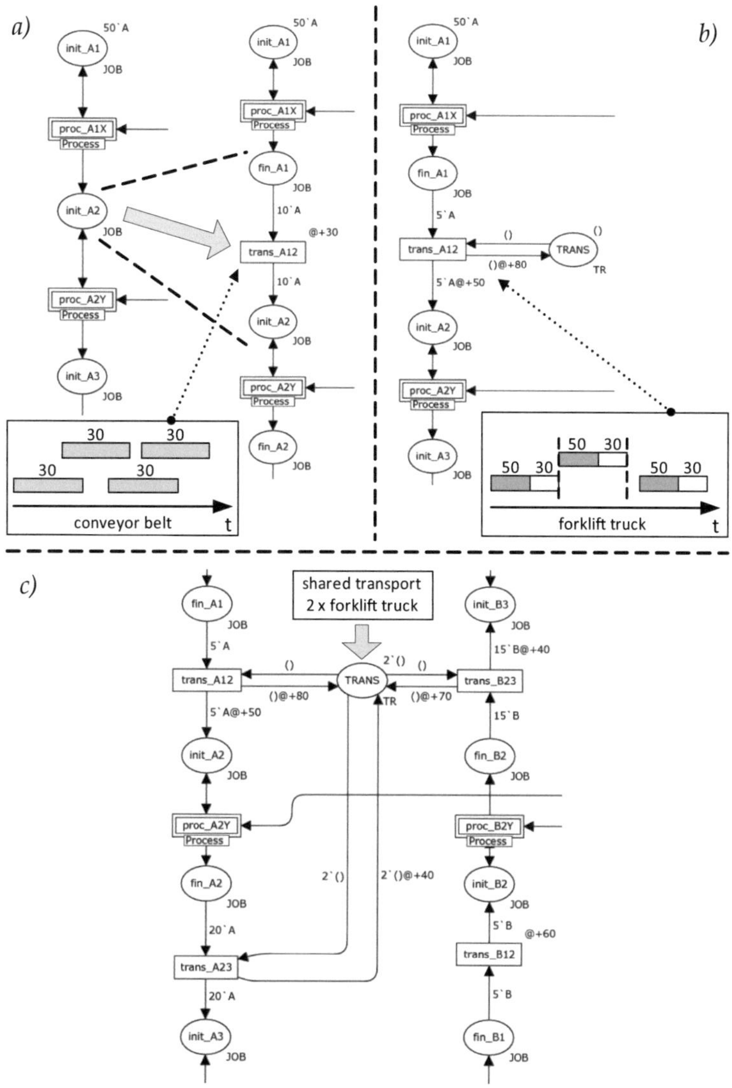

Fig. 8. Modelling of transport: a) transport delay, b) dedicated transport, c) shared transport.

4.5 Limiting structures

It is quite simple to model basic limits characteristic of production systems by Petri nets. The limit of release time (not-earlier-than restriction) is often used. It can be modelled as shown in the figure 9a. The additional place *rel_A2Z* with the initial marking ()@140 causes that the operation represented by the transition *proc_A2Z* cannot be executed ahead of time 140.

If a place models a buffer of material waiting for processing, its capacity sometimes needs to be limited. It can be done using so-called *anti-place*, that is a place connected with the same transitions as an original place but with arcs directed inversely. The total number of tokens in a place and its anti-place remains constant, so the initial marking of anti-place determines maximal capacity of the related place, provided that this place is initially empty. For example (fig. 9b), the place *init_A2* has the capacity limited to 20, because of the anti-place *buf_A2Z*.

An anti-place does not have to relate to a one place in a net structure, it can encompass a greater subnet. An example is presented in the figure 9c. The anti-place *ctr* causes that the total number of tokens in the route between the transitions *proc_B1Z1*/*proc_B1Z2* and *proc_B2Y*, so in the places *fin_B1* and *init_B2* jointly, cannot be greater than 15.

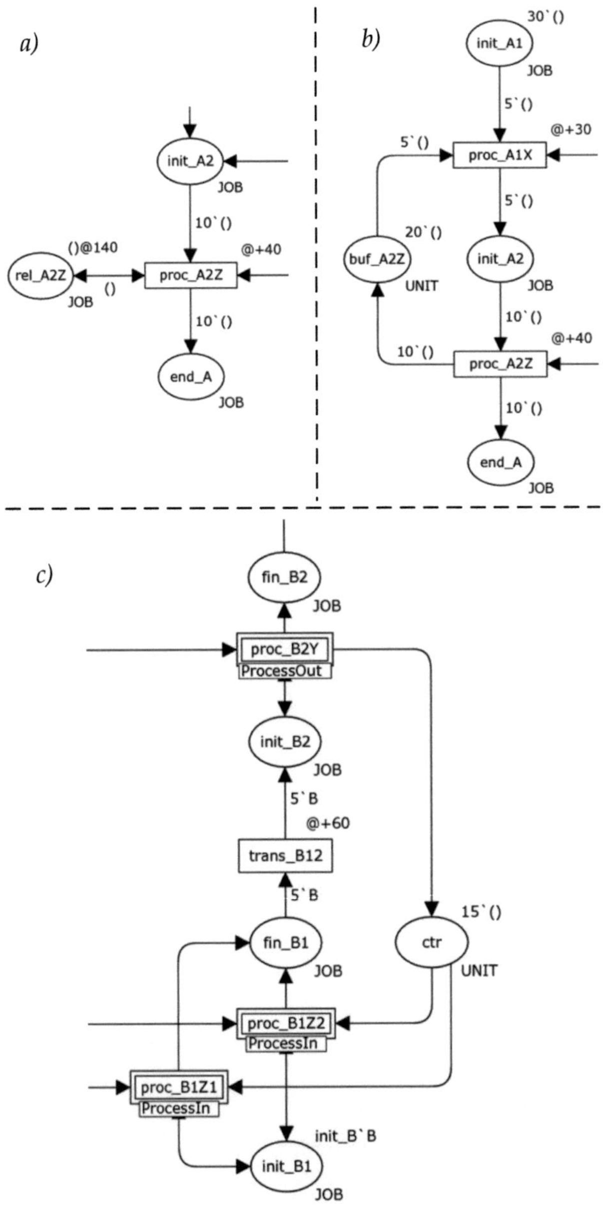

Fig. 9. Limiting structures: a) not-earlier-than restriction, b) place capacity limit, c) subnet capacity limit.

5. Design of scheduling module for screw factory

The introduced rules of production system modelling by the HTCPN have been applied in practice. A prototype scheduling module has been created for a medium size screw factory. This is the part of a larger project that strives for implementing an innovative *flexible manufacturing system* (Christo & Cardeira, 2007). Hardware and software infrastructure for the project has been created and the real-time monitoring subsystem has been developed so far (Żabiński & Mączka, 2010). It has given a chance to have very up-to-date information for a scheduling module.

The general concept of the solution is that all needed information about the state and the structure of the production system is acquired from the main computer system of the factory. This information is used for generating the HTCPN structure that reflects present state of the production. The generated net structure is simulated to foresee next events in the system. The simulation is driven by a priority based controller that makes choices if a decision is needed which of machines or jobs to select for processing. The set of events registered during the simulation is transformed to a production schedule and it is presented in the form of a Gantt chart. All the procedure of data acquiring, net generation and simulation as well as results presentation is executed automatically on demand of a planist.

5.1 Production profile

The production structure in the factory has a form of a flexible job shop system. Beside the standard relations and constraints characteristic of FJS, there are specific features of the system:

1. Batch processing is applied. Batch sizes are determined by sizes of containers for screws.
2. For each machine an individual calendar of availability is defined.
3. For each operation a not-earlier-than restriction is given.
4. For each triple (machine, previous job, next job) a setup time is defined.
5. There are virtually only non-delay production sequences executed.
6. For some of jobs due dates are specified.

There are about hundred machines taken into account in scheduling. A typical schedule includes a few hundred jobs and each job consists of two to ten operations.

5.2 Software structure

The general software structure of the developed flexible manufacturing system is presented in the figure 10. The *main system logic server* unit represents a fundamental business logic of the system that currently realizes tasks of real-time monitoring, data acquisition and generating of statistics. The part that performs the scheduling algorithm consists of the *CPN scheduler and "what-if" generator server/client application* blocks.

The scheduling module acquires data from the main system logic server. This data can be divided into two parts:

1. A set of information that defines the current production state registered by the real-time monitoring subsystem. This information specifies real states of machines (working, stopped, setup, etc.) and jobs (processing, stopped, cancelled, etc.).

2. A set of information that has been defined regardless of the production state. It represents jobs that have to be processed in the future and all their parameters (an amount, a batch size, a technological route, processing times, etc.).

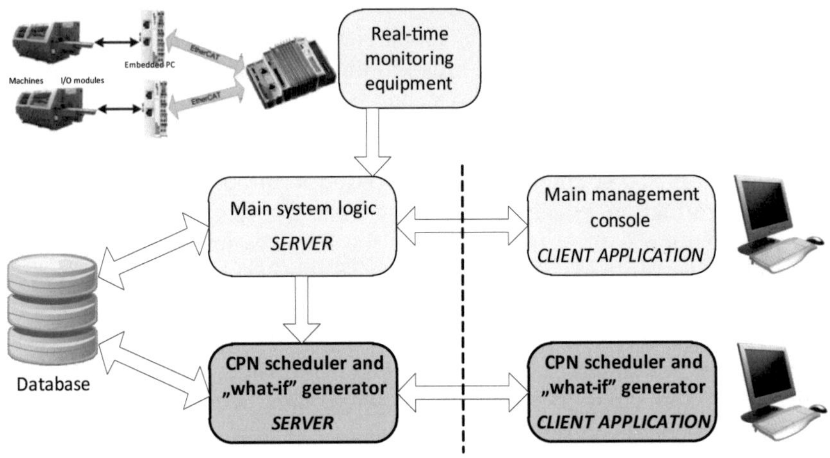

Fig. 10. The general structure of production monitoring/planning/scheduling system.

Both the sets of information are needed for generating up-to-date schedules. The scheduling module has also an access to a database where it stores its own persistent data.

In the figure 11, the inside view of the *CPN scheduler and "what-if" generator sever* module has been shown. The block A (*generator of HTCPN structure*) generates HTCPN structure using data taken from the *main system logic server*. The generated net structure has a form of a set of connected objects. The objects are coded in Java programming language. Any object is either a place or a transition. Arc expressions are included in the transition objects. Net construction rules are similar to that discussed in the subheading 4.3.

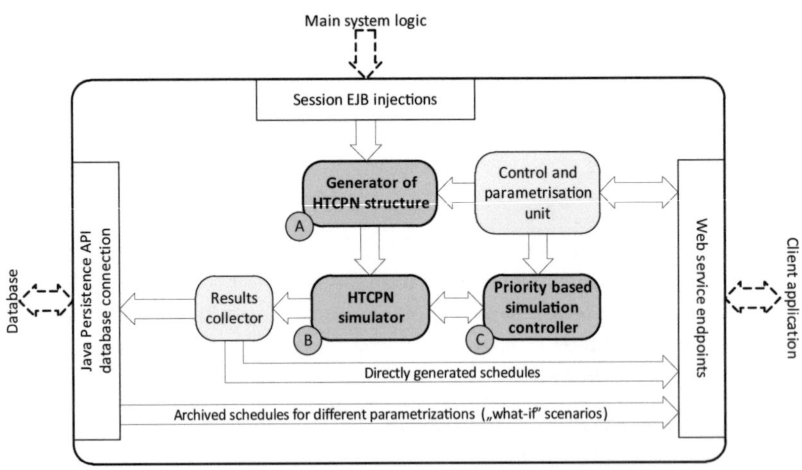

Fig. 11. The inside view of the CPN scheduler and "what-if" generator sever module.

The block B (*HTCPN simulator*) simulates the net structure generated by the block A. If a decision is needed during a simulation, the block C (*priority based simulation controller*) is invoked. The *control and parametrisation unit* block uses data received from the client panel to control details of the blocks A and C execution, for example, due dates of jobs can be changed. Events generated during a simulation are registered by the *result collector* block. The set of results is written to the database and it is sent to the client panel where the results are presented on a Gantt chart. It is also possible to take from the database schedules previously generated for different data sets and parametrisations. These schedules can be compared to discover how changes of parameters affect results ("what-if" scenarios).

5.3 HTCPN simulator

The *HTCPN simulator* block (fig. 11, block B) has been developed specially for the presented scheduling module. It is possible to use the HTCPN simulation engine that is part of *CPN Tools* but this possibility has not been chosen. It has been caused by the several reasons:

1. A net structure has to be generated dynamically for a current state of production. The *CPN Tool* engine supports simulation of a previously edited net diagram.
2. The dedicated simulation block is faster, because it does not have to implement all elements of HTCPN formalism. The elements of the formalism that are not used in the models can be omitted.
3. A special connection between the simulator and the priority based simulation controller is used in the dedicated implementation.

The block diagram of the Petri net simulation algorithm is shown in the figure 12. At the start of any step of the simulation (fig. 12, section A) an enable time of all transitions is verified and there are created two lists. The list *enabledTrans* includes transitions with the enabled time not greater than the value of the global clock and the list *minTimeTrans* includes transitions with the smallest value of the enable time. The smallest value of the enable time is kept in the variable *minTime*. If the value of *minTime* is not valid (*minTime* = *Long.MAX_VALUE*), there are no enabled transitions and the simulation stops (fig. 12, element B). If the value of *minTime* is greater than the value of the global clock, the global clock has to be increased (fig. 12, section C).

A special solution has been used to connect the *HTCPN simulator* with the *priority based simulation controller*. Transitions have been divided into two sets: the set of *decision* transitions and the set of *non-decision* transitions. Only *decision* transitions can be in a conflict in a one simulation step, that is, if one of these transitions will be executed, some others can lost enabling. That conflict represents a scheduling decision. Remaining transitions are consider as *non-decision*. For example, only *proc* transitions in the diagram from the figure 5 have the *decision* type, other transitions are *non-decision* (*setChoose, calCtr, removeOp, setTime*). If any *non-decision* transition is enabled in a given simulation step, it is executed immediately (fig. 12, element D). If a step has only *decision* transitions enabled, the *priority based simulation controller* (called also *optimizer*) selects one of them to the execution (fig. 12, section E).

It should be noted that implemented rules of the simulation and the transition selection do not violate the HTCPN formalism. It is permissible during the simulation to inspect enabled transitions and to select them in a deterministic way. It is also possible in *CPN Tools*.

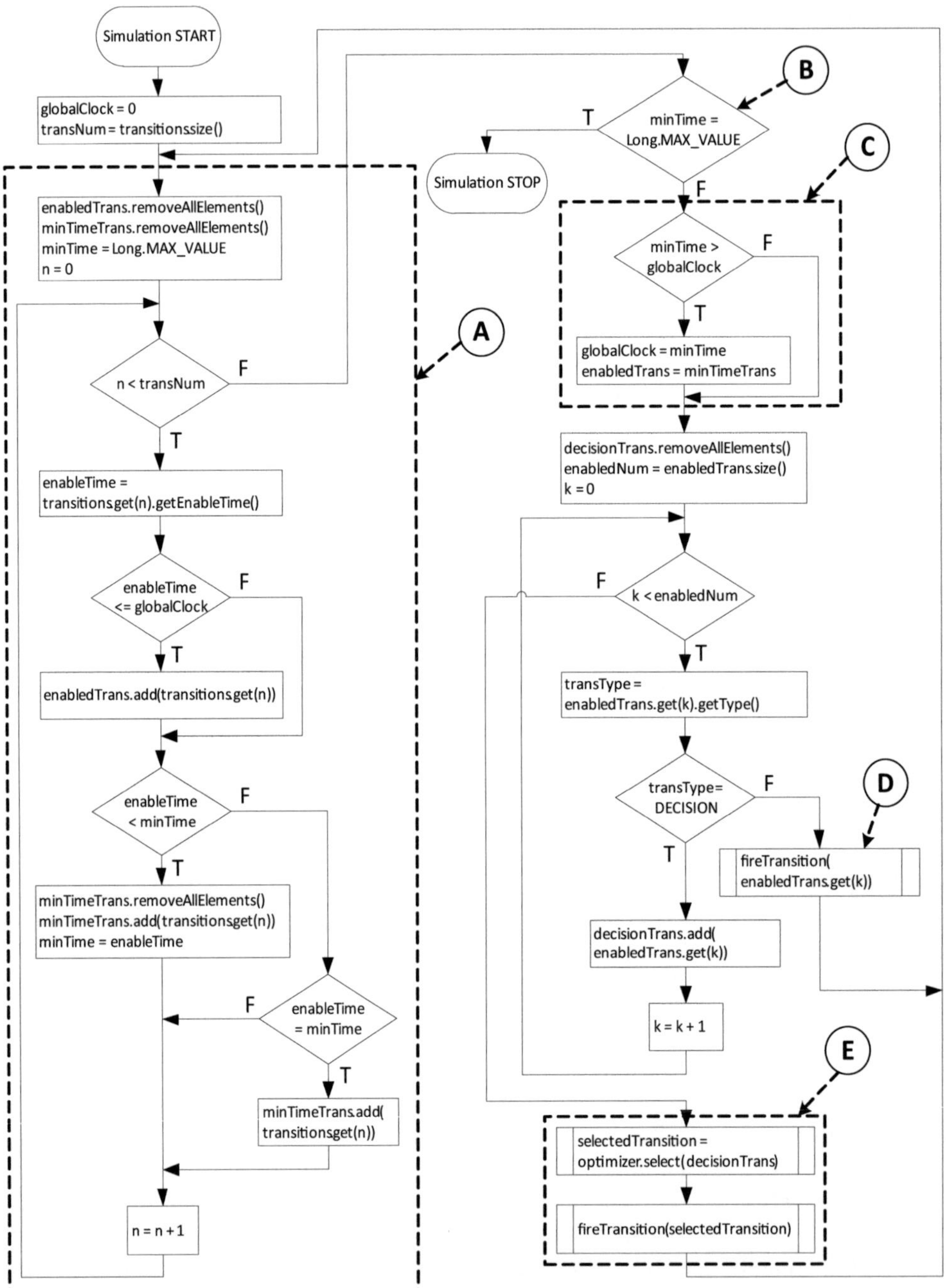

Fig. 12. The block diagram of the Petri net simulation algorithm.

5.4 Priority rules of scheduling

If a conflict between jobs or machines is present, a choice is made on the basis of priority rules. Two main parameters are taken into consideration when priorities are determined:

1. Any of jobs can have assigned a hard (constant) priority. Hard priorities are represented by natural numbers that have to be assigned in sequence.
2. Any of jobs can have assigned a due date.

The rules that control the job selection are hierarchically ordered:

1. The hard priorities are the most important.
2. If there are no operations with the hard priorities (at a given simulation step), the value of the critical ratio is taken under consideration, but the critical ratio value has to be below the threshold fixed as the algorithm parameter. The jobs that have no due date assigned are omitted in this selection step.
3. If the critical ratio of all waiting operations is above the fixed threshold, the operation from a job that has been started earliest is scheduled first. It is done to balance execution time of started jobs.
4. If there are no operations from started jobs, i.e. each waiting operation is the first operation of a job, release dates of jobs are considered in the FIFO order.

The hard priorities should not be overuse because it disrupts functionality of dynamic scheduling rules.

It is rare but possible in the considered production system that many machines wait for a selected job. In that case, the machine having a minimal processing time is chosen.

6. Scheduling module in practice

The main window of the scheduling client application is presented in the figure 13. In the bottom left corner of the window there is a list of generated schedules. A planist can remove and create new schedules freely. If any of the schedules from the list is selected, a table with schedule parameters appears in the bottom panel of the window and a Gantt chart appears in the upper panel. Only the operations that are planned to start later than the moment defined by *freeze time* parameter can be reordered in a scheduling process.

It is possible to only refresh a schedule (*Refresh* button) or to initiate the full scheduling process (*Schedule* button). When the schedule is refreshing, sequence relations between operations are not modified, only newest information about the production progress is taken from the real-time monitoring subsystem to update start and end dates of operations. In the refreshing mode, the *priority based simulation controller* (fig. 11) reconstructs operation order saved in a database. In the full scheduling process, operations are ordered on the basis of dynamically determining priorities that have been presented in the subheading 5.4.

A planist can manually modify the generated schedule by moving some operations using the interactive Gantt chart in drag-and-drop fashion. The HTCPN simulator controls modifications and refuses the changes that violate sequence constraints. The optional values of hard priorities and due dates (fig. 13) that influence final priorities can be set by a planist and can be different for each variant of a schedule.

Various schedules can differ in parameters of jobs (hard priorities, due dates) and also in the changes manually done by a planist. It is possible to compare different schedules and to choose the most favourable. The window *scheduling results* (fig. 14) is used for results comparison.

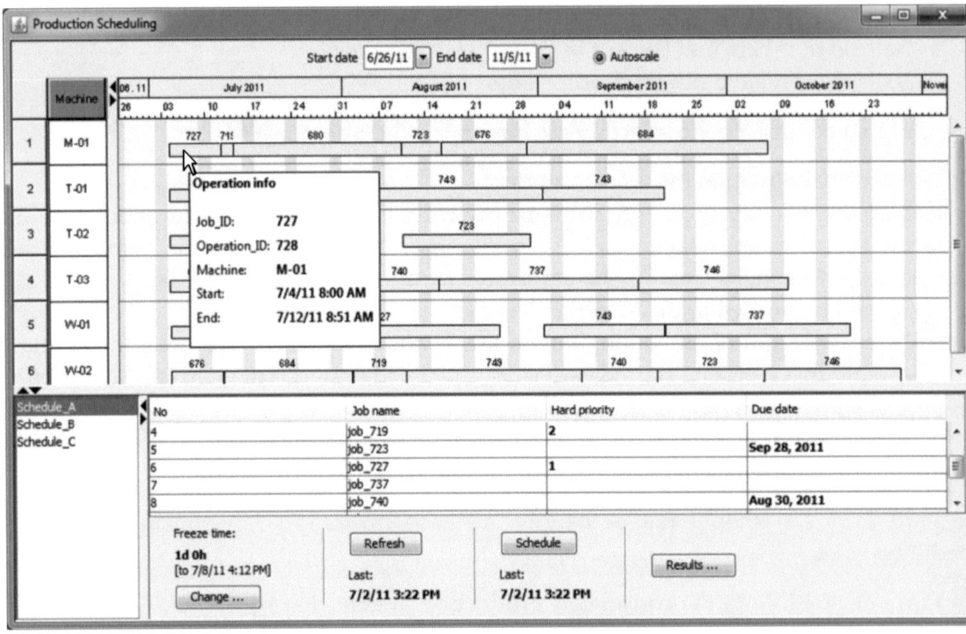

Fig. 13. The main window of the scheduling client application.

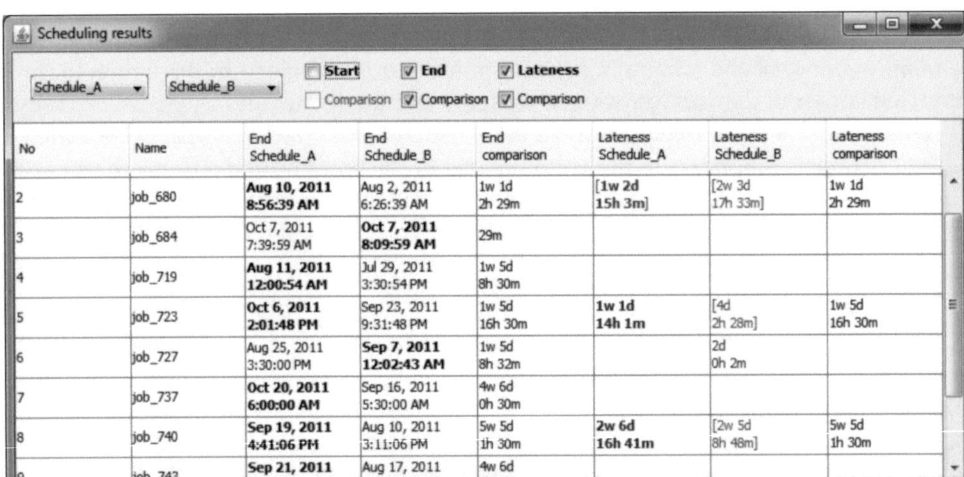

Fig. 14. The result window of the scheduling client application.

Any pair of the generated schedules can be selected and compared. A lateness value is computed for the jobs that have defined a due date. If a job is not late, a negative value of a lateness is presented in square brackets. For example (fig. 14), *job_723* is late over one week in the *Schedule_A*, but it is scheduled about four days in advance in the *Schedule_B*.

7. Conclusion

In the chapter, the hierarchical timed coloured Petri net (HTCPN) formalism has been presented as a powerful and flexible tool for modelling and simulation of complex production systems. It has been proposed how to represent different elements and behaviours of production systems with the use of the HTCPN formalism and how to build complete models of production systems. The main contribution of the work is the presentation of the set of modelling rules together with the evidence that these are useful in practice. In contrast to the majority of other works, basic concepts of differing jobs by token colours and representation of processing time in a timed net have been moved to a background. Instead, high level structures have been taken into consideration that represent batch processing, machine calendars, transport, etc.

It is possible to point two main kinds of application of the HTCPN formalism in production modelling, simulation and scheduling:

1. Models can be built and simulated only in *CPN Tools* or a similar environment. The design process of a net model is fast and convenient in the such case. This approach can be used for finding and verification of production system properties.
2. The HTCPN formalism can be used for creating a new algorithm of simulation and scheduling.

There are, of course, advantages and disadvantages of using of the HTCPN formalism. The HTCPN formalism combines structural modelling and textual programming. So, typical for discrete event systems mechanisms as synchronisation, resources sharing or mutual exclusion can be represented structurally, while others as complex numerical computation can be coded in the textual programming language. The great advantage is also that the time representation is included in the formalism. On the other hand, complex HTCPN models are difficult to the formal analysis and the state space inspection. The next disadvantage is that the HTCPN net model can have relatively high computational complexity of simulation, especially if no optimisation is used.

The future work related to the scheduling system in the screw factory includes:

1. Design of a more optimised block of the HTCPN simulator, because not all possibilities of optimisation have been implemented in the present solution.
2. Verification and adjusting priority rules used for scheduling.
3. Verification whether it is possible to combine developed rules of HTCPN modelling with process mining procedures (van der Aalst, 2010) to build net models on the basis of system logs.

In the future work, it is also planned to use advanced optimisation procedures, as meta-heuristic algorithms, for control a net model simulation, instead of simply priority rules.

8. References

Aized, T. (2010). Petri Net as a Manufacturing System Scheduling Tool, In: *Advances in Petri Net Theory and Applications*, Aized, pp. 43-58, InTech, ISBN 978-953-307-108-4, Croatia

Bożek, A. & Żabiński, T. (2010). Colored timed Petri Nets as tool of off-line simulating for intelligent manufacturing systems. *Przegląd Elektrotechniczny (Electrical Review)*, Vol.2010, No.9, (September 2010), pp. 101-105, ISSN 0033-2097

Brucker, P. (2007). *Scheduling Algorithms*, Springer-Verlag, ISBN 978-3-540-69515-8, Berlin Heidelberg

Camurri, A.; Franchi, P. & Gandolfo F. (1991). A timed colored Petri nets approach to process scheduling, *Proceedings of CompEuro '91, Advanced Computer Technology,* ISBN 0-8186-2141-9, Bologna Italy, May 1991

Christo, C. & Cardeira, C. (2007). Trends in Intelligent Manufacturing Systems, *Proceedings of* IEEE *International Symposium on Industrial Electronics,* ISBN 978-1-4244-0754-5, Vigo, June 2007

CPN Tools Homepage. (n.d.)., 30.06.2011, Available from http://cpntools.org

Jensen, K. & Kristensen, M. L. (2009). *Coloured Petri Nets: Modelling and Validation of Concurrent Systems,* Springer-Verlag, ISBN 978-3-642-00283-0, Berlin Heidelberg

Pinedo, M. L. (2008). *Scheduling: Theory, Algorithms, and Systems,* Springer, ISBN 978-0-387-78934-7, New York

Tuncel, G. & Bayhan, G. M. (2007). Applications of Petri nets in production scheduling: a review. *International Journal of Advanced Manufacturing Technology,* Vol.34, No.7-8, (October 2007), pp. 762-773, ISSN 0268-3768

Van der Aalst, W. M. P. (2010). Process Discovery: Capturing the Invisible. *Computational Intelligence Magazine, IEEE,* Vol.5, No.1, (February 2010), pp. 28-41, ISSN 1556-603X

Zhang, H.; Gu, M. & Song, X. (2008). Modeling and Analysis of Real-life Job Shop Scheduling Problems by Petri nets, *Proceedings of 41st Annual Simulation Symposium,* ISBN 978-0-7695-3143-4, Ottawa, April 2008

Żabiński, T. & Mączka, T. (2010). Human System Interface for Manufacturing Control - Industrial Implementation, *Proceedings of 3rd International Conference on Human System Interaction HIS,* ISBN 978-1-4244-7561-2, Rzeszów Poland, May 2010